Henry Ullyett

Rambles of a Naturalist round Folkestone

With Occasional Papers on the Fauna and Flora of the District

Henry Ullyett

Rambles of a Naturalist round Folkestone
With Occasional Papers on the Fauna and Flora of the District

ISBN/EAN: 9783337025809

Printed in Europe, USA, Canada, Australia, Japan

Cover: Foto ©berggeist007 / pixelio.de

More available books at **www.hansebooks.com**

RAMBLES OF A NATURALIST ROUND FOLKESTONE,

WITH OCCASIONAL PAPERS

ON THE

FAUNA AND FLORA OF THE DISTRICT,

TO WHICH ARE ADDED

LISTS OF PLANTS, LEPIDOPTERA, BIRDS,

AND

LAND AND FRESHWATER SHELLS,

BY

HENRY ULLYETT, B.Sc., F.R.G.S.,

Honorary Secretary to the Folkestone Natural History Society.

" For O but the world is fair, is fair,
And O but the world is sweet !
I will out in the gold of the blossoming mould,
And sit at the Master's feet."

FOLKESTONE:

J. ENGLISH, STEAM PRINTER, HIGH STREET, AND ALL BOOKSELLERS.

1880.

J. ENGLISH, STEAM PRINTER, FOLKESTONE.

CONTENTS.

I. RAMBLES ROUND FOLKESTONE. PAGE.

  1. The Lower Sandgate Road   1

  2. Sugar Loaf Hill and Holy Well   ...   12

  3. Castle Hill   20

  4. The Warren   27

  5. Our Snakes   43

  6. East Wear Bay—Geological   49

II. OCCASIONAL PAPERS.

  1. On the Study of Natural History ...   55

  2. Winter Work   63

  3. The Blood Beetle   71

  4. The Slow-worm   76

  5. The Large Green Grasshopper   ...   80

  6. On Primroses   85

  7. The Puss Moth   93

  8. The Clouded Yellow Butterfly   ...   105

  9. On Violets   116

III. LISTS.

  1. The Plants of Folkestone   129

  2. Butterflies   139

  3. Moths   140

  4. Birds   150

  5. Land and Freshwater Shells   ...   156

PREFACE.

This little volume, which I would fain hope, may prove an acceptable *Vade-mecum* to all lovers of Nature who may either reside in the district, or may honour us with occasional visits, has been published in consequence of the numerous enquiries made every season by our Naturalist-Visitors, as to what there is to be found, and where to search for it. Folkestone is, by this time, a household word in the mouth of every naturalist; few places can compete with it in the variety and abundance of its "Common Objects." This is shown conclusively, I think, in the lists appended, wherein are enumerated about 700 species of Plants, and the same number of species of Lepidoptera.

In the first part I have endeavoured to show, in language as free from scientific technicalities as possible, what natural objects of interest may be noticed in ordinary rambles in our own immediate neighbourhood.

The second part consists of a selection of papers, some of which have been read before our Natural History Society, while the others are reprints from *Science Gossip* (by kind permission of the Editor). They all relate to the Folkestone Fauna and Flora.

Part three contains lists of species known to occur here. It will be noticed that in them I have not given any localities. Some of the more interesting are mentioned as they occur in the "Rambles"; all botanists and entomologists will, I know, appreciate my reasons for not going into particulars in all cases, especially when the species may be rare. I shall be very glad, however, at any time, to give what information I may to any naturalist who may think well to honour me with a call.

I am indebted for these lists much more to the endeavours of my friends, than to my own researches, although I can personally vouch for the greater portion of them. In drawing up the list of plants, I have received invaluable assistance from Mr. J. Hanbury, of London, and Mr. G. C. Walton, of Sandgate.

The list of Lepidoptera is founded on that of Dr. Knaggs, published with valuable notes for the Natural History Society in 1870. By the aid of Mr. C. A. Briggs, of Lincoln's Inn, and Mr. W. Purdey, of Folkestone, however, we have been able to certify the occurrence of more than double the number of species recorded in that list.

The list of Birds is by Messrs. F. Tolputt and V. Knight.

And lastly the Land and Freshwater Shells were kindly worked out for me by Mrs. FitzGerald, who has devoted considerable attention to them, and possesses a splendid collection.

To these, and to all other friends, known and unknown, I hereby offer my best thanks for their assistance.

RAMBLES ROUND FOLKESTONE.

———:o:———

I.—THE LOWER SANDGATE ROAD.

———:o:———

The old sandstone cliffs on the one hand, the ever-sounding sea on the other;—who that knows Folkestone does not know the Lower Road? Judging by the numbers that frequent it we should say there are few rambles in the neighbourhood to be compared with it ; be it summer or winter, it is never deserted. What shall be our musings as we wander along the pebbly beach, or stroll up the romantic path cut on the face of the cliff through the miniature groves of Scotch Fir ? Perhaps it will be wiser, certainly more successful, to give our thoughts the rein, and let them carry us whither they will, for at every fresh step we shall gain fresh reminders, and old associations will rush back again on the mind.

These old grey cliffs, what tales they could tell of the days that are gone ! What stirring scenes have once and again been witnessed from their summit ! Time was, when the road at their foot was a rocky sea shore, over which the waves dashed unceasingly in the days when mankind were no less tumultuous than themselves : possibly, nay some will have it,

B

very probably, the woad-stained Briton, and the
hoary Druid watched from thence the proud fleet of
the Romans nearing the shore ; and since then Saxon
and Dane, and Frenchman have in their turn paid
hostile visits and left unwelcome tokens behind.
On that platform, in the days when it stretched
much further seaward than now, was erected the
first Religious House in Britain, presided over by no
less a person than Eanswythe, daughter of Eadbald,
King of Kent, but ages ago it was washed away by
the hungry waves ; and the original parish church
shared the same fate.

Vessels, equipped and manned by our forefathers,
started again and again from below in obedience to
the summons of the Warden of the Cinque Ports,
even up to the time of the Armada, and after ; then
gradually the warlike tendencies died out, and not a
hundred years ago the whole locality was a great
smuggling depôt. They smuggled English guineas
across to France, and they smuggled tobacco, silks,
and spirits back to England, and that so daringly
that the cargoes were openly displayed on the beach.
Clefts and caves abound in the chalk cliffs which
were excavated for hiding places, and the curiously
built old houses in the ancient part of the town, the
intricate passages, and the trap doors, which rouse
the curiosity of the visitors, are thus easily accounted
for. But now quieter times are come, the fishing
smack and the collier, with an occasional cargo of
timber are all that find ingress and egress at the
port, and there is leisure for grander things. To the
Naturalist, who has learnt to

> Look on Nature as a volume
> Ever open to inspection,

this Lower Road must needs be a favourite resort, whatever his tastes, there are treasures here worth the seeking, and mysteries which a lifetime would fail to unravel.

To the *Geologist* it is interesting as the commencement of that escarpment of Greensand which runs round the great Weald district to meet the sea again at Beachy Head. The strata appear to dip five or six degrees to the east, but the true dip is probably N.E.; we may follow these lines past the Harbour until they enter the bed of the sea. No profusion of organic remains is to be detected in them; many of the blocks contain coarse grains of quartz and glauconite, showing that the waters of the old sea were far from tranquil, an opinion borne out by the many evidences of false bedding and cross-stratification presented in the face of the cliff, and wherever a quarry is opened. The rocks belong to the subdivision known as the *Folkestone Beds*; a short distance past the Bathing Establishment the *Sandgate Beds* crop out—a dark mixture of clay and sand, in which we may occasionally find fossil wood much bored and tunnelled by worms.

A little more interest attaches to a limited deposit of Pleistocene age crowning the cliff at the back of the Pavilion Hotel, just below the Battery. In this, some years ago, were found bones of the extinct Mammoth, Rhinoceros, Hippopotamus, Irish Elk, Reindeer, &c., bringing before the mind's eye another picture of the past, older and stranger than any of the historical pictures already noted. It is a freshwater deposit, as shown by some shells found in it by Mr. McKenny Hughes, during a hurried visit.

But where is the river that laid it down? Was it
the stream which cut out the Folkestone valley, the
representative of which now ignobly creeps along
under the Viaduct, and then, hastening out of sight
through the mill and along the back of Tontine
Street, buries itself in the Harbour? Can that be
the impoverished descendant of a river on whose
banks the huge Mammoth found food, in whose
rushy bed the Hippopotamus bathed? Who knows?
And *when* was it all done? Can we tell the ages?*
Certainly at that time, whenever it was, the extinct
elephant was no scarce beast in the locality; its
bones cannot be said to be uncommon, they are now
and then turned up out of the clay in the brickfields.
Part of a fine shoulder blade, now in the Museum,
was disinterred in 1876 near the Cheriton Arch,
close by the spot where, a few months before, some
teeth and bones of the Rhinoceros were found. I
have also a box full of bones belonging to a skeleton
of a very small Mammoth found on Park Farm in
1868. A very slight exercise of the imagination
suffices to picture out the stores of vegetation which
must have covered all this district, as necessary to
support these huge mammals. Like their modern
representatives in Africa, they would require a
luxuriant herbage, and we may fairly compare this
district with their present dwelling place. But they

---

* I am constantly asked by those whose curiosity exceeds
their taste for geology *how many years* ago it is since these
creatures lived. But the geologist reckons not by years,
cannot so reckon, however much he may wish it. His chro-
nology is relative, not absolute, for he has no certain data
on which to base his calculations.

themselves are all gone, they are a vanished race.

"The old order changeth, giving place to the new."

They probably disappeared before the advances of the half naked but more sagacious Celt, or even his predecessors, who with their flint-tipped arrows forced on their destruction. Now they are gone too, and we have the nineteenth century with railroads and telegraphs, electric lights and microphones. *Tempora mutantur.*

And now, what of the botany of the Lower Road? It is truly a storehouse of treasures. From the time when the banks are yellow with Furze blossoms in January to the appearance of the orange berries on the Sea Buckthorn the slopes are a blaze of colour. And chiefly it it is golden. The Furze blossoms have not disappeared before the Kingcups or Butter-cups (whichever name you prefer) spread themselves luxuriantly in all directions. Not one or two species only, for *Ranunculus acris* is there, though sparingly; *R. bulbosus* as usual, is without limit, and *R. repens* rears its head crowned with a corolla an inch and a quarter across, two feet and more above its lowlier neighbours. Then in June, mixing with the dying blossoms of the *Ranunculaceæ* comes the glorious hue of the Bird's-foot Trefoil *(Lotus corniculatus)* and a host of its relatives. In no one spot have I ever seen gathered together so many representatives of the *Leguminosæ*, and among them such good things. I give at the end of this ramble a list of the species found on this road. The botanist will notice par-ticularly *Lathyrus Nissolia*, and *L. pratensis*, *Vicia bithynica*, *Trifolium suffocatum*, and *T. arvense*, He will, I know, want to start at once for *Lathyrus*

Aphaca, but that, like so many good things, belongs
to the days of old; I have never seen it, but Mr.
Mackeson, of Hythe, tells me he gathered it there
many years ago. But *Nissolia* is still to be seen,
though in greatly diminished numbers, for the path
has been cut close by, if not through, the very spot
where I saw it every year. The same place was also
sacred to the Deptford Pink *(Dianthus Armeria)*; I
have never seen them elsewhere. I have not been
able to find either of them since 1877, though I have
been told of a solitary specimen of each gathered in
1879; I would fain believe they are still lingering in
some unknown, and therefore safe nook. The patches
of *V. bithynica* are rather extensive; it is a vetch
which attracts notice at once by the white wings of
the flower, backed by the purple standard. *Trifolium*
and *Medicago* are represented by several species, and
L. pratensis is here in groves. *T. arvense*, the Hare's-
foot Trefoil grows only in one small patch near the
summit above the turnpike gate.

On the slope below the Old Church the Notting-
ham Catchfly *(Silene nutans)* crowds together, and a
small patch of the fragrant Butterbur *(Petasites
fragrans)* appears quite naturalized close by. The
Fennel *(Fœniculum officinale)* betrays its presence by
its odour just behind the Bathing Establishment,
where it is surrounded by tall straggling heads of
Charlock *(Sinapis arvensis)* and its close relation
S. alba. A few years ago I found a grand specimen
of the Dame's Violet *(Hesperis matronalis)* but it has
since disappeared. Beyond the turnpike gate the
cliff face in May and June is covered with the pretty
pink blossoms of Thrift *(Armeria maritima)*, and the

common Stone-crop (*Sedum acre*) flowers along the border of the upper path; we may also find a few specimens of it on the rocks on the opposite side of the road, in which patch of ground flourishes also the Burdock (*Arctium Lappa*) Here too by the side of a stream containing just enough water to keep its channel marshy, I was shown four or five years ago the bright blossoms and leaves of the Bog Pimpernel (*Anagallis tenella*); it was plentiful then, and the specimens I gathered have lost little of their beauty, but the slips and shiftings of the ground seem now to have caused its disappearance. Here and there along the road we find the Tree Mallow (*Lavatera arborea*), the Woody Nightshade (*Solanum dulcamara*), and the Teazel (*Dipsacus sylvestris*). The Brambles are in profusion, but (except in fruiting time) I always fight shy of them, having a wholesome fear of Babington and his forty-nine species. There are the Dewberry (*Rubus cæsius*) and the Blackberry (*R. fruticosus*), that is all I venture to say about them.

One thing, however, the Lower Road lacks, namely trees, they do not seem to get on there at all; the Scotch Fir thrives the best, but even these are sorry specimens, libels on their species to any one who has seen the magnificent trees in some parts of Scotland. Several hundreds have been lately planted on the slopes, the majority of which bid fair to succeed, so that in a few years there will be a great improvement. And while we are speaking of this Fir, let me tell you that it is worth while tapping the blossoms with your stick in the early summer only to see the enormous clouds of yellow pollen which will fill the air. Concerning which pollen, strange tales

were afloat during the summer of 1879. In the
neighbourhood of Eton and some other spots, it
was noted that the ponds of water after rain were
covered with a yellow powder. Forthwith it got noised
abroad that there had been a shower of sulphur,
and it was not reckoned polite by some persons to
guess where it came from ; one good wife in describ-
ing it, said " it smelt awful strong," filling the air
round about with its odour; many good people
guiltless of botanical lore " improved " the occasion,
others (true Englishmen) wrote to the papers about
it. A certain botanist, however, who possessed a
microscope, annihilated (or attempted to annihilate)
all their speculations by declaring the yellow dust to
be merely the pollen of the Scotch Fir. Some one,
hailing from the Emerald Isle, declined the annihila-
tion, and in his turn wrote to the papers, saying that
whatever the powder might prove to be in England,
it was neither more nor less than genuine brimstone
in Ireland, and rashly offered to send some to the
learned botanist. The offer was accepted, and once
again (as I need hardly remind my readers) Ireland
was robbed of her hopes by the Saxon.

I must now add *N.B. Caution to Botanists.* You
will find on some parts of these slopes specimens of
bona fide garden flowers, the seeds of which were
scattered about by a friend of mine in order (he said)
to increase the attractions of the cliff. I know what
the opinions of many botanists concerning such
practices are ; I express none, I only give an explanation which might possibly be needed.

To the *Entomologist* I can only give a few directions.
Many lepidoptera are common enough here, but

there are two or three worth hunting for. Of
Tapinostola Bondii you may take as many as you
want; the larvæ have been found feeding up in the
stems of *Festuca arundinacea*, a grass which is very
abundant here. On the stunted poplars "once upon
a time" Dr. Knaggs found larvæ of the Anchorite
(*Clostera anachoreta*) which duly produced imagos,
the ancestors of most of the specimens now in
English cabinets. I have myself reared *C. curtula*
and *C. reclusa* from caterpillars off these trees.
Notodonta dictæa and *N. ziczac*, two of the Prominent
Moths, are still to be found; the larvæ of the Puss
Moth (*Dicranura vinula*) may be taken by scores,
and *D. bifida* has occurred here. The Satin Moth
(*Liparis salicis*) and the Brown Tail (*L. chrysorrhœa*),
the Poplar and Privet Hawks abound, and an occasional Death's Head turns up, feeding either on the
potatoes in the cultivated patches or on the Garden
Tea-tree which surrounds them. Apropos of this insect let me tell you a tale with a moral. A friend of
mine out in a boat one evening caught a fine specimen which had settled on the sail. He brought it
home and allowed it to wander about the room till
after tea, when he purposed to kill it. Accordingly
we had a chase after it until it settled on my arm as
if appealing to me for protection, little thinking that
I too had a vacant space in my cabinet labelled
A. atropos. On my attempting to seize it, it uttered
its well known "shriek," which sounded so like a
piteous appeal for life that my friend's tender feelings overcame his entomological desires, and he
declared at once it should not be put to death. So
he put it under a glass shade, where in a few days

he found it starved to death. *Moral:*—Poison your
insects at once.

At the flowers of *Silene nutans* you may very
probably in June take *Dianthecia albimacula*, an insect worth pounds three or four years ago in the
market, but now its value is reckoned in shillings.
It unaccountably turned up again in this country
when everybody thought it was gone, and not only
asserted its right as a Britisher, but came in numbers.
Some dozens are caught every year in the Folkestone
district. Will some of the other insects now on the
list of the dear departed ever come back to our land ?
Shall we ever see the Large Copper again ? More
unlikely things have occurred. A few larvæ and
imagos of the Cream-Spot Tiger (*Arctia villica*) are
taken every year, but they get scarcer, so by the bye
do those of *Arctia caja*—Woolly Bears, for which it
is difficult to account. I have reared fine broods of
the Emperor Moth (*Saturnia carpini*) from caterpillars
off these brambles ; how is it there have been none
the last year or two ? But 1879 was a bad year for
larvæ, at any rate in this locality, else generally the
bushes in the hollow past the turnpike gate have
been swarming with them.

The only butterflies worth noticing here are the
Clouded Yellows (*Colias Edusa* and *C. Hyale* with a few
Helice) which come occasionally in their unaccountably independent fashion. *Melitea Cinxia* is also
said to have formerly fed on these slopes.

In conclusion, if you, gentle reader, should belong
to that unhappy section of mankind known as the
non-naturalists, I may yet say you will enjoy this
ramble. You may wander along and allow your

thoughts to rove whither they will, and that is very pleasant occasionally. It is pleasant to throw one's self down in one of the mossy hollows just beyond the Half-way Rocks, and let the benign influences of Nature herself have full sway. There, as the last rays of the sun tinge the waters, and the distant horizon becomes more and more obscure, a pleasing sadness will come over you, full of longings which you cannot utter; the white sails before you fading away into dimness and uncertainty will carry your thoughts on to the "Land beyond the Sea," into which you hope to sail when the twilight of Life comes on, and winds and storms shall have become things of the past.

LIST OF THE LEGUMINOSÆ FOUND IN THE LOWER ROAD.

Ulex europæus.	Vicia hirsuta.
Ononis arvensis.	,, Cracca.
,, campestris.	,, bithynica.
Medicago sativa.	,, sativa
,, maculata.	Lathyrus Aphaca.
,, lupulina.	,, Nissolia.
Trifolium pratense.	,, pratensis.
,, arvense.	Hippocrepis comosa.
,, scabrum.	Onobrychis sativa.
,, repens.	Lotus corniculatus.
,, procumbens.	Anthyllis vulneraria.

II.—SUGAR-LOAF HILL AND HOLY WELL.
————:o:————

As the visitor, approaching Folkestone by rail, passes over the Viaduct, the most prominent object on the landward side is undoubtedly " Sugar-Loaf Hill," projecting from the long escarpment as if thrusting itself on his notice.   Let us on some quiet summer evening pay it a visit.

We will start from the foot of Grace Hill, going up the opposite slope and ascending the " Bull-dog Steps."   Only a few years ago, and this road was a narrow undrained lane affording a passage past the Mill, with a spring of beautiful clear cold water at the foot of a flight of fragmentary and dislocated steps—a spring noted for its hygienic virtues certainly for the last two hundred years, for do not the town records give evidence of the care with which it was protected ?   But this, with many other interesting evidences of antiquity, has lately made way for what I suppose we must call improvements, but we get them at the expense of the picturesque, and in the eye of the antiquarian and the naturalist such improvements are not always regarded as worth the sacrifice.   But it must ever be so it would seem ; the age is one of utilitarianism, and apparently that is incompatible with the picturesque.

At the top of the present flight of steps we were formerly able to get a good view of the Railway Viaduct, a light airy structure by Brunel, the middle

arch standing 120 feet above the stream that drains
the valley. Often have I paused at this spot, just
at set of sun to enjoy the prospect; there was then
a splendidly luxuriant specimen of the Horse Chestnut
tree on the right, the only one of any size in Folkestone, it has since been cut down to make room for
a shop. Leading thence by the banks of the stream
up to the Viaduct was a fine row of tall poplars, for
them too we mourn, unconsoled even by the pleasant
row of comfortable houses occupying their place.
We keep grumbling at the lack of trees in Folkestone,
but all the time we go on cutting down what few we
have. Very grateful to the eye was that row of
poplars leading up to the arches, and beyond all was
the background of hills fading away to the west in
the purple twilight. It is difficult now to find any
spot on this side from which to get a good view, and
streets are springing up so rapidly that if we allow
a few months to elapse we can hardly find our way
about old familiar spots.

On passing through the arches we seem at once to
emerge into a new country, the whole scene changes
and so does its character. There before us stretches
the bold chalk escarpment of the *North Downs* with
its hollows and gently swelling sides; and across
the grassy slopes falls the evening sunlight in long
golden gleams. It looks like some grand old earth
fortification of the giants
　　　" To sentinel enchanted land ;"
one feels an instinctive desire to go up and see what
lies on the other side. The contrast between plain
and highland seems to give great pleasure to the
mind. I shall never forget the feeling of pleasant

astonishment I once experienced when, after a few
hours' ramble among the Buckinghamshire chalk
hills, I turned a corner of the road and suddenly
found myself on the edge of the escarpment, the flat
plain lying three or four hundred feet below me, and
stretching away for miles in the distance. So again,
after climbing up Wansfell near Ambleside for 1500
feet one July evening, did an exclamation of wonder
burst from my companion on seeing the grand low-
land on the other side.

But *revenons nos moutons*, we are in the suburb of
Foord and not in Westmoreland. And there in
front of us, prominently outlined against the clear
sky stands the hill up which we intend to wend our
way. We go on through the rapidly growing hamlet,
and past the old sham ruins the destruction of
which has already commenced. Inside, so the old
guide books tell us, a few years ago, was a mineral
spring, even dignified as chalybeate, but no trace of
it can be detected now. On, past the new church
of St. John Baptist, missing the interesting old
cottages on the right which have long since given
place to brick and Folkestone cement; the very
stream itself which bordered the road a little further
on has been diverted, I suppose because of the pro-
jected street through the old Pavilion Gardens. On
the left here is the only spot in the neighbourhood
where the Sand Martins used to build, perhaps I
ought to say *attempted* to build, for I question whether
they were ever allowed to succeed, so strong is the
juvenile desire for ornithological research. But here
again the pickaxe and shovel are at work, rapidly
removing temptation out of the way of the birds.

All this, ten years ago, was in a very deed a country lane—a very pleasant stroll for those whose walks were limited in extent.

We now turn down a short lane leaving the two laundries on our left and enter the borders of Park Farm. There are just sufficient traces of antiquity here to make us wish there were more. Part of the moat still remains, which is said to have surrounded the old Manor House of the Middle Ages, the site of which is now occupied by a small cottage. The plot of ground thus partially surrounded by water must originally have been much larger; it is about 120 feet square.

On, over a couple of meadows and we reach the gate at the foot of the Sugar-Loaf. Here let us rest a few minutes, and look at the scene we have left behind. From no place (except perhaps from the summit) could we get a better or prettier view of Folkestone. There is the Viaduct, showing now to advantage as it gracefully spans the valley; beyond it the town, lit up by the evening sun; further still the sea, quiet and calm, for there is scarcely a breeze to ruffle it—a perfect picture of *rest*, for the vessels themselves seem at this distance to have no motion at all. And between this and ourselves the spacious domain of Park Farm, alas! no longer a park. Once upon a time—that time when we are apt to fancy (is it all fancy?) everything was more perfect than now—once upon a time all this was a beautifully wooded district, and now the only complaint against our town is as I have said, that there are no trees. Little more than 300 years ago the estate was cleared by order of Henry Herdson, then Lord

of the Manor; the small patches of wood still exist-
ing near Cheriton and Newington are probably the
remaining trees of this once noble park. In the
words of Mr. Mackie "Peace be to his ashes, but I
wish he had let it alone;" why he did it we cannot
say, probably from circumstances over which he had
lost control; since then Folkestone has been treeless.
May success attend the efforts recently made to
restore them in the town itself.

From our resting place an easy path leads up the
side of the hill, bordered by a hedge in which flourish
luxuriantly wild Roses, the White and Black Bryony,
the Mealy Guelder Rose, Dogwood and Privet. At
the top on the right is a little copse charitably spared
(let us hope) by the farmer, rich in Orchids, Leguminosæ, and Knapweeds. It was here that
in 1870 I saw my first and only Golden
Oriole, and a splendid bird it was. A gorgeous
visitor like this is soon noted, and as it went down
into the hollows I saw the bird catchers on the alert
gazing with wistful eyes and ready nets. I am happy
in being able to record their disappointment. Now
let us mount to the summit; have you not already
wondered at the curious shape of the hill, and what
the meaning is of the evidently artificial formation
of the top? Well, in all probability, it is a *barrow*,
an old British or Saxon burial place; but as to who
it was, or how long ago the burial took place, I can
tell you nothing. Men talk of a bloody engagement
in the year of grace 456, somewhere in the neighbourhood, between the Saxons under Hengist, and
the Britons led by Vortimer, who vainly sought to
make up for his father's lack of patriotism; it has

even been said that this was the sepulchre of the
British prince. Who knows?

> " I cannot say how the truth may be,
> I say the tale as 'twas said to me."

But they are pleasant things to chat about, these old
legends. I give you this for what it is worth.
Anyhow, it is very delightful to look around us from
such a vantage ground,—the Farm, the Railway and
the Town in the foreground, and beyond, " the white
sails of ships "—on the left the old cliffs of the
Warren, to which we mean to devote another ramble,
and here, down the steep slope on the right is a
curiously shapèd recess in the escarpment, the lower
portion of which is covered by the quiet waters of
Holy Well, fed by two or three streams issuing from
the base of the chalk. Why " Holy " Well I cannot
say, we must again fall back on tradition ; it is also
" Lady Well," and about a mile from it over the
brow is " Lady Wood," but whether any connexion
exists or ever has existed between the two is all unknown. We might indeed, give full and free scope
to our imagination without incurring any severe
criticism. The hand of man is evident all around,
the artificial bankings and smoothings of the surface,
the made road now covered with grass leading up to
the top, speak plainly enough of his work. May we
not think of an old Religious House standing here,
whose inmates drank of the cool clear waters of these
springs in the days when Christianity was less of a
" civilized heathenism " and more of a grand reality
in the world than it is now ;—when men instead of
saying " I believe," *felt* it, *lived* it, and acted it out—
when " their religion " as Carlyle says " was *the great*

c

fact about them ?" Tradition supports us thus far,
for the Well is said to have been a stopping place for
pilgrims landing on the coast close by on a visit to
the shrine of the murdered Becket at Canterbury.
Truly the spot seems well suited for such a house ;
let us descend, and seating ourselves on that little
grassy platform in the middle, indulge our thoughts
still further. Here the inmates would be wholly and
completely shut out from the busy world, all its
traffic, all its disappointments. You can now perhaps
see one or two of the mansions at the new West End,
but in those days Old Folkestone was out of sight
here, snugly hidden beneath the cliffs, not even the
smoke of it would be visible. And so, free from all
distraction, from all sounds save those sent by GOD
Himself—the voices of the birds and the trees, and
ever and anon the swell of the ocean borne aloft on
the breeze above the old forest, they could give themselves up to that contemplation which finds no con-
genial home among the busy haunts of men.

And now, what have we here in the way of botanical treasures ? Nothing very extraordinary, but
much to give pleasure. The slope we have just
descended is covered in June with the Spotted and
Pyramidal orchids (*Orchis maculata* and *O. pyramidalis*),
later on we find a few stray specimens of the Bee
Orchis *(Ophrys apifera)*, and in the autumn the Lady's
Tresses *(Spiranthes autumnalis)*. The waters themselves are well nigh hidden by the abundance of
Ranunculaceæ, Pondweeds, Reeds, and Rushes.
The banks round about will yield in their season at
least three species of violets—the Hairy Violet (*Viola
hirta*) and two forms of the Wood Violet (*V. Riviniana*

and *V. Reichenbachiana*)* with Cowslips and Primroses; the marshy spot over which the waters find an outlet give us shining masses of the Golden Saxifrage (*Chrysosplenium oppositifolium*) and spear forests of the Yellow Iris (*Iris Pseudacorus*). In the little gully beyond is the Twayblade (*Listera ovata*), not very common at Folkestone, but found also in Lady Wood, and further along large patches of what you might at this distance take to be Lilies of the Valley, but which on a nearer approach betray themselves as Garlic (*Allium ursinum*).

A very pleasant path leads from the gate above, along the slope to the foot of Cæsar's Camp, but as the shades of night are already beginning to close in round us, we must postpone any further explorations till our next ramble.

* Along the south side of the hedge, running at the base of the hill the specimens of *V. Riviniana* are remarkably fine, growing seven or eight inches in height.

III.—CASTLE HILL.

————:o:————

We may reach this very prominent natural feature
of the district by a pleasant path from Holy Well
along the foot of the escarpment. Or, for a change,
suppose we go up the Dover Road, and then turn off
to the left, just above the limekiln. We shall then
get a good panoramic view of many miles of both
land and sea (possibly a sight of Boulogne Cathedral,
and that without a glass), and also a good look-down
on the curiously shaped summit of Sugar-Loaf Hill.
A path across the fields from the turnpike gate,
through a district known as "Gibraltar" will take us
on to *Castle Hill*, or as it is also called, *Cæsar's Camp*.
The view obtainable here is even more extensive than
that from the Sugar-Loaf, reaching from Dover Pier
on the left to Dungeness on the right. We appear
to be in a district rich with "memories of the past;"
the hand of man is visible all along the brink of the
hills—a tumulus to our left, another here to our
right, while the banks are artificially sloped along
almost the whole face. One would think there must
have been many a struggle here between the early
invaders and those who fought for the "humble
cottage in Britain." Our interest just now, however,
centres in the hill itself. Folkestone is said to have
possessed a Pharos in Roman times, but while that
at Dover still remains, every trace of ours is gone;

several antiquarians have thought it stood probably
on this very spot. It does not seem very likely that
it would have been at such a distance from the sea.
But Lambarde says that in his time " there were yet
extant to the eie the ruined walls of an antient fortification, which for the height thereof might serve for a
watch tower." However, there are no traces whatever now of such a building, whatever it may have
been. Intrenchments there are, outer and inner,
plainly traceable, but antiquarians cannot agree
about those who dug them. Why it is commonly
called *Cæsar's Camp* no one can tell, for there are no
certain records in writing that the Romans ever en-
camped in the neighbourhood. Some will have it
the works are Celtic ; excavations were made here a
short time ago under the superintendence of General
Lane Fox, and escarpments were laid bare, evidently
military work, and a deep well on the summit was
partially emptied ; beyond however a few doubtful
pieces of pottery and a small specimen of archi-
tecture that was Norman, I believe little was found.
General Fox is however to read a paper giving an
account of it shortly, and then we shall know what
conclusions he has been able to draw from the ex-
cavation.

We will leave then the historical associations, and
follow out a pursuit at the present moment more con-
genial to our minds.

The slopes of the hill make one extensive carpet
of flowers all through the summer. We have here a
grand metropolis of orchids, not only the commoner
ones, the Early Purple (*Orchis mascula*), the Spotted
(*O. maculata*), and the Green-winged (*O. Morio*), but

Bee orchids literally by hundreds, and the Late
Spider *(Ophrys arachnites)* not at all rare among
them. The Pyramidal *(Orchis pyramidalis)* and the
Sweet-Scented *(Gymnadenia conopsea)* grow by
thousands, and I know of one specimen at least of
the Man Orchis *(Aceras anthropophora)* found here.
Two species of Rest-harrow *(Ononis arvensis* and *O.
spinosa)* are common, while the Milkwort *(Polygala
vulgaris)* and the Wild Thyme *(T. Serpyllum)* are
almost as plentiful as the grass itself. All along the
hedgebank at the foot we find the parasitic Broom-
rapes *(Orobanche)*.

And for the Entomologist there are *Adonis, Alsus,*
and *Corydon* in their appointed seasons, *Semele,
Cardui, Rhamni, Hyale,* and *Edusa* with stray *Helice.*
One of the grandest entomological sights I ever witnessed was on the hill there beyond the waterworks.
It was a summer evening, bird and flower alike were
preparing for rest, when I saw as I stood at the top,
the whole face of the slope covered with innumerable
specimens of *Adonis* resting with outspread wings
on the grass, their rich blue throwing back the rays
of the declining sun. It afforded me intense pleasure,
and no little astonishment, for there had been a
heavy and prolonged thunderstorm the evening
before, and I had expected to find few of these small
butterflies about, at any rate in a respectable condition. How they managed to hide themselves at all
in that downpour I was at a loss to imagine, but how
they succeeded in also preserving their resplendent
colour deepened the mystery. But there they were,
and I shall never forget the picture.

Among the moths we get *Mi* and *Glyphica,*

little *Ænea*, and all three of the Foresters (*Statices*, *Globulariæ*, and *Geryon*). Further along the hills westward several years ago I took many specimens of the Five-spotted Burnet (*Z. trifolii*) flying about with *Z. filipendulæ*, but I have never seen it since, either there or in any other part of our district. All along these slopes we find in October the handsome larva of the Fox Moth (*Bombyx rubi*) in great abundance, the perfect insect appearing in May. Only once have I been fortunate enough to see it on the wing, and then, of course, I had not my net. It only flies for an hour or so towards the end of the afternoon, and then only the males are about; the females hide themselves in the long grass, and the only way to find them is to track the males patiently until they drop down. I took three males in my hat, and they are still my only specimens. Why should it be so difficult to rear this insect? The caterpillar hybernates, and gets on very well during the winter, but in captivity ninety-nine out of every hundred invariably die in the spring, only a few entomologists have succeeded in rearing them. Along with my persevering friend, Mr. Blackall, I have hunted for the larvæ in April but we have not found above a dozen altogether, and those we did find died, mostly from previous attacks by parasitic diptera. We then searched later on for cocoons among the long grass, furze and brambles, and *two* crowned our efforts from one of which Mr. B. was fortunate enough to rear a female. *Bombyx rubi* is in fact, a troublesome insect to get hold of. I strongly suspect that the caterpillar spins up at the end of its hybernation without going abroad much to eat, otherwise we should most certainly find plenty of them in the early spring. They are very prolific, but they have evidently numerous enemies, as we found many fragments of cocoons.

The little nook by the side of the hill is known as the *Cherry Gardens* and contains the Folkestone Waterworks. A few years ago we should certainly have gone down and obtained some refreshment, for the air on these hills is appetizing, but now the cherries have departed, and the place is "private." Let us follow the path and stroll through the short lane and along the meadows homeward. There is one little object of interest here, if you are anything of an antiquarian, and that is the miraculous stream which flows uphill. In the field next to that through which our path runs, we may find a tiny little aqueduct through which one stream is carried, and under which runs a second. This aqueduct (so says tradition) was built by St. Eanswythe, daughter of King Eadbald (see page 2), to allow of the stream being conducted to her religious house for the supply of the inmates. There was a slight impediment in the way however, namely, that the elevation of the nunnery was greater than that of the source of the spring; but she who "restored the blinde," and "forbade certaine ravenous birdes the countrey," "drewe the water over the hills and rockes against Nature," and the enterprise was successful. And even now this same stream supplies a large pond on the Bayle, near which spot the nunnery stood. If you will take your stand anywhere along the course of this stream, and look up it, you may easily fancy the source to be much lower than the spot where you are. Scientific men of the present day tell us that the last thing in the world to which we are justified in trusting, is the evidence of our own senses. I give you this caution and there I leave the matter.

As we follow our proper path we may notice a small pond on the left, which I mention here for the sake of the beautiful sight it presents in May and June. Its surface is then covered with the large white blossoms of one of the Water Crowfoots (*Ranunculus aquatilis.*) Once only have I seen a similar sight that surpassed it; it was in June, 1879 when I was walking along the canal side from Hythe towards Lympne. For two miles without a break, the water was one mass of blossoms. Once or twice every year I come across the fields for the sake of the view of snow white blossoms mantling this pond; on a small scale it is to me what the "host of golden daffodils" was to Wordsworth,

> "And oft when on my couch I lie,
> In vacant or in pensive mood,
> They flash upon that inward eye
> Which is the bliss of solitude."

There are at least one or two moments in most men's lives productive of pictures like these—scenes from the Distant or the Past which flash on the eye of the mind and fill the heart with intense delight. One or two for each of us. Mine are but such as may fall to the lot of any ordinary person; but one stands out supreme—the moment when I stood gazing on those magnificent falls of Cora Linn and Bonnington in that beautiful park near Lanark—gazing in silent worship, filled with awe, mingled with gratitude that it should have been given to me to behold such

beauty, fresh almost it seemed from the hand of GOD.
Another, nearer home, as I lay resting on one of the
steep slopes on the cliff between St. Margaret's Bay
and Dover Castle ; the bright sun lighting up both
land and water, the sea like glass, and studded with
sails, the regular thudding of the paddles of the
steamers reaching the ear from a distance of two or
three miles—a scene of enchantment, all so quiet, to
be looked at only in silence and enjoyed.

Along the banks of this little stream where the
trees are thickest, it is worth the entomologist's while
to devote a night or two to sugaring. The three
Red Underwings *(Catocala nupta, C. sponsa* and
C. promissa) have been taken here.

IV.—THE WARREN.

————:o:————

" It's O my heart, my heart!
To be out in the sun and sing;
To sing and shout in the fields about,
In the balm and blossoming.
Sing loud, O bird in the tree,
O bird, sing loud in the sky;
And honey bees blacken the clover beds,
There are none of you glad as I.
The leaves laugh low in the wind,
Laugh low with the wind at play;
And the odorous call of the flowers all
Entices my soul away."

Mr. Stanley, in the exciting description of his
" Journey through the Dark Continent," tells us
how, in one spot where he knew he could ramble
about in perfect safety, he began his explorations
with all a boy's ardour—how he gave free rein to the
impulse to run, jump, spring, and climb, to roll
about on the leaves or ruins of branch and bark, to
laugh and sing—how, in short, he was for an hour
or two in the enjoyment of such bliss as can be
enjoyed only in communion with Nature, and in the
bright fresh days of Spring. If you, gentle reader,
would wish for such pleasure as was evidenced by
these private demonstrations, it is attainable on the
Warren, which I counsel you strongly to visit, not
once, nor twice, but often. If you wish to make a
good day of it, I would recommend you to start from
St. Peter's Church, just above the Harbour, and
travel along the edge of East Cliff. You will get

here the full benefit of the sea breezes, and a
pleasant view of the Harbour and Railway Station.
You can watch the fleet of fishing boats starting on
their journey, and swinging round the Pier Head,
and the Boulogne boat taking a more graceful sweep
as she brings her load of possibly pale and uncom-
fortable travellers to the landing stage, where they
may once more feel themselves steady on terra
firma.

This field, simply known as *East Cliff*, has long
been a kind of common playground, though not I
believe professedly so ; but landlords and farmers are
very indulgent in our neighbourhood. My London
friends tell me they enjoy coming to Folkestone so
much because they "can go about everywhere" without meeting horrible " Caution " boards—a liberty
unknown in the country suburbs of the great city.
It seems probable now, however, that the freedom so
long enjoyed on this cliff will be somewhat restricted,
if not entirely taken away ; the Genius of bricks and
mortar has seized upon the place for his own, and it
is doomed to be built over. Let us enjoy it while
we may.

See, as you walk along the edge, what destruction
occurs year after year by the breaking away of the
cliff, a destruction, I cannot help thinking it is quite
possible to stay. People call it the ravages of the
sea ; that is not quite correct, there must be some
other agent at work, for the top of the cliff wears
away faster than the bottom. Far more destructive
than the action of the waves is that of the fresh
water which abounds in this part. The soil and sub-
soil are of no great depth, and are very porous ;

immediately beneath is the **Gault clay**, which, by stopping the percolation of the water causes it to collect in underground pools and rivulets; these burst out in the face of the cliff, as you may see in a dozen places, and rush down to the beach below, carrying with them an enormous amount of sediment. A proper system of draining, together with the construction of a sea-wall (which might consist simply of the blocks scattered about below) would certainly check, if it did not altogether prevent the waste; but it must be a full, and not a half finished system. It would appear that the greatest falls have occurred just at those places where the drain pipes terminate, the water discharged has worn back the face of the cliff, then the end of the pipe has broken off, and this has gone on alternately until, within the last ten years twenty or thirty yards have, in some places been taken off from these gardens. I do not know on what tenure the land may be held, but I should imagine at an annually decreasing rental.

About the geology of the cliffs we will have a chat in another chapter; it will be sufficient here to state that we are walking chiefly over the Gault, but that it is covered with a deposit of diluvium some three or four feet deep, in which are thickly embedded numbers of flints, all very sharp and angular, the remnants of the chalk which once evidently covered the whole plain. If we went farther back from the face of the cliff towards the escarpment we should soon detect an increasing thickness of rainwash from the slopes, composed of chalk mud with small lumps of chalk scattered through it.

At the termination of these garden patches we

come to the *Warren* itself, as picturesque a spot as
you may find anywhere in the country, a broken and
confused stretch of ground for about two miles along
the base of the cliffs, an undercliff itself, mostly
natural, partly artificial from excavations and
embankments made to allow the railway to pass
through it. It would be a puzzle to any geologist
to arrange its strata, if indeed it have any, for
gault and chalk, sand and masses of flint are
in inextricable confusion. Yet this very confusion gives to the place its beauty; there are not fifty
yards of level ground throughout its whole extent;
hence we get a number of lazy ponds in quiet sunny
hollows, giving us throughout the summer months
glorious pictures of still life. Who that lives at
Folkestone, who that visits it, does not love the
Warren? There are three Martello Towers; from
No. 3 you will get the best view, an artist's view of
the whole,—beautiful in its contour, rich in its
contrasts of colour—the bold sweep of East Wear
Bay, backed by the massive cliffs of chalk, diversified
with verdure, and capped here and there with patches
of red clay and sand—Tertiary Drift.

Let us describe this western end first. It is by no
means comparable to the further portion, but it is
not without its beauty; it is perhaps the most broken
part of all, and the most unstable, changing every
winter. The luxuriant growth of rushes and horse-
tails (*Equisetum maximum*, and *E. arvense*) tell you
how damp it is as you look over it, but as you follow
the path you see for yourself the pools in almost
every hollow, and the water travelling from one to
the other over newly exposed clay. Like the cliff

along which we have passel, it is being destroyed
by freshwater action, the effect of which is to cause
enormous masses to be thrown downwards towards
the sea, there to be devoured by the waves. A walk
along the beach from the Harbour towards Dover in
the early Spring shows the amount of damage ; I
have seen thousands of tons of débris lying about,
the effect of one winter's work, but by the end of the
Summer it has all been washed away again. Not
much more than a dozen years ago there stood a
cottage and garden on this end of the Warren ; in
fact, I believe it was once a gentleman's residence,
with a road down which one could drive to it, but
notice of ejectment was served upon the last tenant
by Nature herself in the shape of successive openings
in the walls, and he wisely quitted it. None too
soon ; now you will only find traces of its walls, and
perhaps one or two lingering specimens of its garden
productions. Decidedly the Warren is not the place
to build upon.

You will want to know about the botany of this
portion of the Warren. Well, there are Brambles
of various species known only to " Splitters " of the
botanical ranks ; climbing over them occasionally some
magnificent specimens of the Tufted Vetch (*Vicia*
Cracca) with its long blue clusters of blossoms ;
then there is Eyebright (*Euphrasia officinalis*), Cath-
artic Flax (*Linum catharticum*), Restharrow (*Ononis*
arvensis) and Lungwort (*Pulmonaria officinalis*) ;
the Centaury (*Erythræa Centaurium* and *E. pulchella*),
and Yellow-wort (*Chlora perfoliata*) with the Gentian
(*Gentiana campestris*), and here and there scattered
tufts of Henbane (*Hyoscyamus niger*) with its half

inviting, half forbidding blossoms, sickly odour, and clammy touch. The Bee Orchis is here sparingly, the Pyramidal Orchis is plentiful, while in the Spring months the Greenwinged Orchis (*O. Morio*) the Spotted Orchis (*O. maculata*) and the Early Purple (*O. mascula*) are abundant. Tall Teazels raise their heads in company with the Ox-tongue (*Helminthia echioides*), clumps of Hemp Agrimony (*Eupatorium cannabinum*), and various species of Thistle including *Carlina vulgaris*. Each season of the year has its own characteristic flora here, always worth coming to see.

*   *   *   *   *   *   *

The ordinary road to the Warren is across the railway leading down to the Harbour, and through the Folly Fields, The first field on the right after crossing the line was long known by the name of "Chapel Field," though no one could say why. But in 1872 some excavations were made there, and the foundations of what had evidently been a church were laid open, they were very massive and strong, and many old Roman tiles were among them. Several skeletons were found, some around the walls, some inside, all lay with their hands across the breast. A curious circular foundation was also exposed close by, but it was apparently unconnected with the building. It was a great pity that the Kent Archæological Society did not pay it a visit, or depute some competent person to do so; it is too late now, for most of the work was pulled to pieces and the stones were devoted to utilitarian purposes. On the left are some brick fields, and the recently established works of the Folkestone Cement Company. Here again the workmen met with the remains of antiquity in the form of a Roman building with its hypocaust, together with pieces of pottery and a few coins. When we get over the next field, past Tower No. 2, we are on the Warren again. Let us go down over the broken ground to this large pond, which is only two or three years old, as you might guess from the bramble and hawthorn trees withered and dead in the midst of the waters like memories of the past. The hollow was formed when the collapse of the tunnel occurred in January, 1877. The place was altogether a scene of ruin and disorder after that event, one small pond was literally turned inside out, and its bed, full of shells of Pond Snails (*Limnæa peregra* and *L. stagnalis*) is now an elevation. Great cracks, some a couple of feet wide and four or five deep, were left in the ground, in fact it seemed as if the Warren had been visited by an earthquake. And such, on a small scale, I suppose we may say it was. This again was the result of the action of fresh water accumulated during a long wet season, and pent up until it exerted its well-known strength and completely wrecked the place. For many long weeks afterwards a stream ran down either side of the railway until it was all drained away. Well, this pond is, as I said, one of its results.

At all periods of the year, and at every hour of the day and night the Warren repays a visit. But to me it is prettiest in Spring and early Summer, doubtless by contrast with its wintry aspect— the days when the milder weather begins to fill the mind with longings after

> "The fall of waters, and the song of birds."

I have just been looking at some notes of a visit

D

paid in June, 1879, to this pond. It was a delightful
spring day (every one knows that there was no summer during that year), and I wanted a few specimens
of the Holly Blue *(Lycæna argiolus)* and the early
Skippers (*Hesperiidæ*), but I saw none ; the only
lepidopterous creatures that entered my net were a
battered green Hair-streak (*Thecla rubi*), and that
very great rarity known as *Plusia gamma*, or the
Silver Y. But the first hour was very delightful.
I got down by the pond and it was teeming with
life, both plant and animal.   Shoals of tadpoles made
the water black all round the borders, there must
have been thousands of them ; the whirligig beetles
and others of that ilk were moving in mazy dances
over the surface, while through the clear water could
be seen the efts lying at the bottom, or sluggishly
coming up to breathe.   Dragon Flies of every size
and hue, the big yellow or blue *Libellula depressa*, and
the airy Demoiselles of richest azure and red, were
there flitting through the air and resting now and
then on the brambles at the side.   It was just one
of those happy moments in a naturalist's life when
he feels what Wordsworth well describes as *the happi-
ness of simply being alive*, when he stands in pleasur-
able wonderment at the exuberance of life around
him.

> " The cowslip startles in meadows green,
>   The buttercup catches the sun in its chalice ;
> And there's never a leaf or blade too mean
>   To be some happy creature's palace."

I noted as a fog came creeping from the east over
the hills what a sudden change came over the Dragon
Flies, how inert they became, you could pick them
off the leaves with your fingers, but when the sun

was out you could get nowhere near them. The Water Crowfoot (*Ranunculus aquatilis*) held up its tiny white blossoms, and the starry heads of *Callitriche verna* studded the surface of the waters, the Lesser Spearwort (*Ranunculus Flammula*) grew thickly round the banks, and close by were masses of the sky-blue flowers of the Brooklime (*Veronica Beccabunga*). I lay upon the bank enjoying it all while the splashing of the tide only a few yards away filled me with that intense longing

"Break, break, break
On thy cold gray stones, O Sea!
And I would that my tongue could utter,
The thoughts that arise in me."

Just over the ridge there on the seaward side you will find another pond filling up a long narrow deep ridge, and called the *Long Pond*. About this spot the shrubs are very luxuriant, the Wild Rose and the Sweetbriar, the Dogwood, the Elder and the two Guelder Roses with Babingtonian Brambles. The blossoms of the Water Plantain (*Alisma Plantago*) crowd above the surface of the water, and over the bushes along the edge creeps the Woody Nightshade (*Solanum Dulcamara*) or Bitter-sweet, with its lurid warning flowers. They are very dangerous these Nightshade plants, just when most beautiful, that is, in the Autumn. I remember some years ago a boy eating some of the tempting berries, and getting frightfully delirious all that night; the physician quite believed he had eaten of the Deadly Nightshade, (*Atropa Belladonna*) from the peculiar symptoms. He got all right again in a few days. I do not know that *Belladonna* occurs anywhere in the district; we have the other three species of the family *Solanaceæ*,

the Black Nightshade *(S. nigrum)* occurring on road
sides, and in waste places. *Lycium Barbarum* seems
naturalized, but the Thorn Apple (*Datura Stramonium*)
has not appeared. I should think it probable that
Belladonna may exist in some of the woods.

By the side of this pond, years ago, I found my
first specimen of the rare Spider Orchis (*Ophrys
arachnites*), but it has vanished again, and though
the Bee is not at all uncommon, I doubt if the Late
Spider occurs now in the Warren. The Early Spider
(*O. aranifera*) is plentiful. From hence all along the
seaward side of the railway the ground is excessively
rough and broken from the results of yearly landslips,
but it is picturesquely clothed with shrubs, notable
among which is the Sea Buckthorn (*Hippophae
rhamnoides*) plentiful here as along the Lower
Sandgate Road. All over the chalky slopes grows
the Yellow Horned Poppy (*Glaucium luteum*), and in
one or two spots *Papaver somniferum.* The Rose-bay
Willow-herb (*Epilobium angustifolium*) fills at least
one hollow, and the slopes are gay with the tall
spikes of the Viper's Bugloss (*Echium vulgare*) rising
amidst masses of Mignonette (*Reseda lutea*), Weld
(*R. lutcola*), and the Wood Sage (*Teucrium Scorodonia*);
the Purple Iris (*Iris fœtidissima*) is very abundant,
and less so the Hound's Tongue (*Cynoglossum
officinale*).

To the entomologist I may say that out of the
grass in June and July we may rouse the little *Venilia
maculata* or a male of the Clouded Buff (*Euthemonia
russula*); on the docks there is a good chance of *S.
chrysidiforme*; *Mi* and *Glyphica* are spinning about
among the herbage, while the scarlet wings of the

Burnet (*Zygæna filipendulæ*) and the Cinnabar (*Euchelia Jacobeæ*) flash in the sun. Why does not its relative the Scarlet Tiger (*Callimorpha Dominula*) occur here too? Nay more, why *will* it not occur, when every provision is made for it? At St. Margaret's Bay on the other side of the South Foreland, it swarms by thousands, feeding on every possible kind of plant; everyone of those plants is abundant on the Warren, many scores of caterpillars and imagos have been turned loose there, but no *Dominula* has ever been bred from them. Why not?

A little earlier, and we should find a few specimens of the exquisite little Lace-border (*Acidalia ornata*). Moths and Butterflies and their larvæ abound; among the latter the Marbled White (*Arge Galathea*), the Grayling (*Hipparchia Semele*), the Dark Green Fritillary (*Argynnis Aglaia*), a few of the Large Tortoiseshell (*Vanessa polychloros*) with an occasional Queen of Spain, a Camberwell Beauty, or a Bath White. In the years when it pleases *Edusa* to appear it swarms all over this locality.

Many a pleasant hunt have I had here after various larvæ, and I remember the time when I pounced eagerly upon stragglers of the Oak Eggar and the Drinker, and rarer specimens of the Emperor Moth. Time and space would fail to tell you of all that might be done on the Warren in the insect line; I hope some competent hand may yet be found to devote a volume to its entomology. I will just give you a list of a few of the " good things " that may be taken here :—On the flowers of *Silene nutans* a few of D. *albimacula;* flying around the bushes for about an hour at dusk *Acidalia strigilata ;* at sugar the Sword-

grass (*Calocampa vetusta*), the Red Underwing
(*Catocala nupta*), and a possible *Leucania albipuncta*;
Heliothis marginata visits the *Echium* in company
with *Chœrocampa Porcellus*—both are abundant;
Aplasta ononaria has been taken once or twice, and
so also has *Deiopeia pulchella*; the Privet, the Poplar,
and the Convolvulus Hawk Moths occur, *D. lineata*
has been taken several times, and of the Humming
Bird Moth (*M. stellatarum*) and *Sesia ichneumoniformis*
you can always find specimens in their season.

That portion of the *Warren* on the inner side of
the railway has been styled " Little Switzerland. "
Not so rough and broken up as the seaward part,
it is nevertheless a well diversified tract of ground,
consisting of minature plains, table lands, and solitary
hillocks, which in spring time almost dazzle the eyes
with the countless number of primroses scattered
over them—

Stars that in earth's firmament do shine.

All through the summer months there is a rapid succession of flowers, making the Warren a favourite
resort for children. I do not know that I could give
a better picture of it than is afforded by the following
extract from my diary :—

" *July 30th, 1879.*—Weather intensely hot. Ram-
bled on till I came to Rakemere Pond. Beyond
all contradiction this is the prettiest spot in the whole
of the Warren, in fact I doubt whether many equally
picturesque places are to be found in the country.
The massive cliffs at the back, streaked with green
and red, the hills all around thickly clothed with trees,
bushes, and flowers, the quiet reedy waters below—
all combine to form a picture which many a time and

oft has been transferred to canvass. As I sat on the summit of one of the hills and looked on the scene at my feet, I could not help saying " How we should enjoy it and value it all if it were but fifty or a hundred miles away." Why do we neglect the thousand and one beauties and wonders close around us, and fly to others simply because they are far away ? However, I did enjoy the panorama before me, and the profusion with which Nature had scattered her wealth around. It came into my mind to count the species of flowers just within my reach where I sat, and this is the result :—

Gromwell.	Eyebright.
Bugle.	Squinanceywort.
Thyme.	Pyramidal Orchis.
Wood Spurge.	Cathartic Flax.
Centaury.	Bird's-foot Trefoil.
Red Clover.	Yellow Bedstraw.
Carrot	White Bedstraw.
Rock Rose.	Hop Trefoil.
Plantain.	Milkwort (white & blue).
Yellow-wort.	

Nineteen in all, any one of which I could gather without leaving my seat. And there, with the Wild Roses blossoming around me, the Privet still in bloom but tinged with decay, the scent of the Sweetbriar perfuming the air, the sea covered with haze from the heat, the only sound the cooing of the Wood Pigeon, what could I do but gaze and offer up that " worship of the silent sort " which prompts the lover of Nature perhaps beyond all other men.

Then I went down to the pond itself and spent a quarter of an hour watching, and listening to the many curious and strange sounds that broke the stillness. Reeds and tall rushes seven or eight feet high filled up one half of the pond, and round the

edge was a thick growth of plants, among which
were the Forget-me-nots, Mints, Watercress, Brook-
weed, Water Plantain, White Water Bedstraw, and,
under the trees opposite, the Enchanter's Nightshade.
Dragon Flies sported about in ferocious glee over the
water; I watched one of the larger species (*Æshna*)
a long time pursuing and snapping up insects in the
air. The muscular strength of these creatures (and
indeed of all insects) must be enormous, when we
consider the time they remain on the wing; they are
able to fly in any direction without turning and are
said to match the swallow in speed. Occasionally
a second of the same species dashed across, and,
whether it was play or deadly combat I cannot tell,
but there was an aerial chase, and now and again
they closed with such force that both fell into the
water together and struggled on the surface, then
they rose and the vanquished individual fled.

Among the odd sounds which kept coming from
the reeds I heard a short sharp barking noise such
as a young dog might make. For some time I tried
in vain to discover whence and from what it came,
but it ceased when I made a slight noise. I passed
out of sight round a bush, and presently heard it
afresh and on working cautiously round got full view
of a Moorhen swimming quickly and unconcernedly
across the open water. It paddled about with quick
short jerks of its white-edged pert-looking tail, constantly uttering its peculiar cry, then it went into
the reeds again. Just as I was preparing to start a
second came in sight following the track of the
former, and differing from its partner only in being
more soberly coloured. For five or six minutes I

watched them swimming about and listened to their monotonous duet; the cry of the female is not quite so canine in character as that of the male, but slightly more musical, a crinking sound, putting one in mind of that made by the frog, only much louder.

The best view of the Warren as a whole—a bird's-eye view in fact, is to be obtained from the top of the cliffs. It is not an easy task to mount them, but still it is possible. I would recommend you, however, instead to take an early walk some morning before breakfast up the Dover Road, whence you will easily find your way. There is a path all along the edge bordered with flowers, notably the small golden crosses of the Ladies' Bedstraw (*Galium rerum*) on which you may find feeding in September or October the caterpillar of the Humming Bird Moth, and where you would like to find (though I fear you would be disappointed in the search) that of *Deilephila galii*.

I witnessed a very curious phenomenon from this path one evening in September. A fog had blown up from the sea to the face of the cliffs, and being thus checked in its progress it became very dense and heaped itself up against them like a horizontal surface of white cloud, through which the Martello Tower and the higher rocks stood out like islands, vanishing and reappearing. Far away to the left we could see this white bank pressing in a sharply defined line against the cliffs, rising like a sea gradually up the steep till in one spot it overflowed, and then in a few minutes it sank quietly down again. All this time we stood quite above the fog, expecting

it to rise and envelope us, but it passed away without
doing so.

You can walk all the way to Dover along this
path, and a most enjoyable ramble it is.   If not too
fatigued you should come back by way of the beach,
that is if time *and tide* permit.   Or should the double
walk be too far you can descend the cliff from the
Coast Guard Cottages at Lydden Spout, and there
below you will find a number of shore plants that are
not common elsewhere in this neighbourhood.
Among them are the Samphire (*Crithmum maritimum*), the Wild Madder (*Rubia peregrina*), the Wild
Cabbage (*Brassica oleracea*), and *Frankœnia lœvis.*
The Wallflower occurs wild on the cliffs on both
sides of Dover.

V.—OUR SNAKES.

————:o:————

Our "Ramble on the Warren" would be considered hardly complete without a few words in reference to its reptiles. Years ago, when I first came to Folkestone, I heard the most marvellous accounts of the snakes to be met with on the Warren —monsters of varied colours, six or seven feet long and proportionately thick. Making the usual allowance for exaggeration, and knowing that the Grass Snake occasionally reaches a length of five feet, I still expected to come across one or two specimens of respectable size. The largest I have seen is in a bottle at the Museum and is scarcely forty inches long.

The Common Grass Snake (*Tropidonotus natrix*) and the Viper or Adder (*Pelias berus*) are tolerably common on the Warren, but the latter is seen far oftener then the former. And yet hundreds of persons frequent the locality and never meet with either. Visitors who have loved to learn this picturesque spot are horrified and disappointed when told of the presence of these creatures. Yet there is not the slightest reason for such feelings, the best evidence of this being the *apparent* rarity of the reptiles ; you may go many times and even search for them without being successful.

Timidity is the characteristic of both species ; at the sound of a footstep they glide noiselessly through

the grass, and seek the cover of the thick bushes or
their subterranean retreats.   Let no one then ever
be deterred from visiting the Warren by the fear of
meeting with either.

The Grass Snake is perfectly harmless, and may
be handled without any possible danger ; it has no
venom, and if it wished to bite, which is highly improbable (I have taken hold of many), its teeth
would scarcely serve the purpose. What *might* prevent your taking up a second specimen would be the
offensive odour which clings to the hand for some
time after. You may easily tell this species by the bright
yellow patches just behind the head, and the absence
of the dark zigzag mark along the back which
characterises its distant relation.

The Viper, I grant, is dangerous to handle, though
not to the extent popularly supposed. It appears to
be still a disputed question whether any one has been
poisoned by a viper, and the poison has *directly*
proved fatal. I say *directly*, for it no doubt may be,
and has been an indirect cause of death ; as for
instance where a person has been bitten in the neck,
and the swelling has produced suffocation. But in
such a case we should hardly say that the man was
fatally poisoned by the viper ; similar consequences
have been known to result from the sting of a wasp.
I have never yet been able to meet with any well
authenticated instance of the poison of *Pelias Berus*
proving fatal in the same way as we should say that
of a Cobra would, that is, purely and simply from its
own nature. Yet almost every country churchyard
has its grave pointed out to children and strangers
as a caution against meddling with snakes and

adders. I remember in particular a churchyard in one of the lonely villages of Norfolk in which was a tombstone ornamented with a sculptured snake with its tail in its mouth forming a ring. Doubtless it was intended as an emblem of eternity, but there it was looked upon as proof positive of the manner of the man's death, and we children used to look at it with awe, while one of our elders related the story of the man gathering wood, when an adder " stung " him, &c., &c.

As there is, no doubt, some residuum of truth even in the wildest legends we may believe that death after the bite of a viper is not an utterly unknown circumstance. The physical constitution of the victim, the state of his health at the time, will no doubt affect the case. Very likely a person of feeble constitution, whose blood was in an impure state, and who chanced to get bitten in the sultry days of July or August, might succumb to the venomous bite ; otherwise I should say not. Such was the case with a man at Folkestone many years ago, he was bitten on the hand and died in a few hours ; but his medical attendant told me that the state of his blood from drink was such that it took very little to kill him. Any instances brought forward on either side of the question must necessarily be interesting. From one or two accounts I have read, and from the following for which I can vouch, it would appear that the venom does not always act in the same way. Mr. Wood mentions a case in which there was intense pain and fever ; in the following instance there was little of either.

I was out entomologising some few years ago,

when I saw a very beautiful viper, about half grown.
Meeting a brother of the net a few minutes afterwards I mentioned it to him. "Of course you killed
it?" said he. "No, I did not, I seldom do."
"Perhaps you were never bitten by one; or else you
always would." No, I was not; I had kept them in
confinement, and I am always shy of killing any
creature I have watched and studied. Upon which
he told me that he was, and the circumstances and
consequences were as follow:—He was out butterfly
hunting, and caught a viper in his net as it was
gliding over the ground. Not knowing then the
difference between a viper and a common snake, he
mistook it for the latter, handled it repeatedly, and
carried it home, where he placed it on a table to watch its
movements. He took it up several times till at last
it turned its head sharply round, and with its fang
punctured the forefinger of the right hand. Still he
took no notice, and continued handling it as before,
though he was careful now to lay hold of it closer to
the head. Shortly, however, he felt a curious
drowsy sensation stealing over him, which his friends
attributed to fancy. But it was not long before he
became seriously ill, his mind wandered, they put
him in bed and sent for a medical man. No olive
oil was applied, and the principal thing given him
was neat brandy in occasional doses. The object of
this was to cause a reaction from the great weakness
which ensued; he felt utterly prostrated, and needed
all that could be given him to restore his physical
strength. He lay in bed a fortnight, no fever
ensued, and more curiously no pain, nothing but
excessive weakness, and, immediately after the bite,

insensibility and delirium. The hand, arm, and side as low as the hip were immensely swollen and almost black ; the two former were frequently bathed in hot water.

Having gone through this little experience, he always made it a rule to kill a viper when he had the opportunity ; not because there was any danger of its attacking anybody, he knew it was a very timid creature, but then " you *might* tread on one." Well, I have trodden on one, but was not bitten, though I stood on it some little time looking round to see where it was, as my friends were calling out to me, but I did not kill it, and it glided off to earth : vipers, like all other created beings, have their allotted work to perform, though we may be ignorant of its nature, and they are neither sufficiently numerous nor wantonly aggressive to warrant our endeavours to exterminate them.

One of my pupils was similarly poisoned in the hand this year (1880). Hand, arm, and side were swollen and black, but there was no fever, and little pain ; he was out at play all the time till the effects were gone. But this was in the month of March, a time when the poison would not be very virulent. Some time ago I had a conversation with a sportsman, in which he mentioned that dogs are occasionally killed by vipers ; he had lost two very valuable ones himself. Young ones generally die but occasionally recover, old ones seldom fall victims. One that he had had some years had been bitten twice when young, but as it grew older it became an adept at killing its foes ; it would spring upon them, all four feet coming down at once, and then with its head up in the air trample them to death.

The Viper is common, not only on the Warren but all over the neighbourhood, it does not seem to be so dependent on water as the Grass Snake. The latter may be seen occasionally swimming in the ponds on the Warren in search of newts or frogs.

Subjoined is a list of the *Reptilia* and *Amphibia* to be found at Folkestone—all on the Warren except the Edible Frog.

REPTILIA.

Grass Snake (*Tropidonotus natrix*)
Viper or Adder (*Pelias berus*)
Slow-worm (*Anguis fragilis*)
Scaly Lizard (*Zootoca vivipara*)

AMPHIBIA.

Common Toad (*Bufo vulgaris*)
Common Frog (*Rana temporaria*)
Edible Frog (*Rana esculenta*)
Large Newt (*Triton cristatus*)
Smooth Newt (*Lissotriton punctatus*)
Palmated Newt (*Lissotriton palmatus*).

VI.—EAST WEAR BAY—GEOLOGICAL.
————:o:————

The geologist possesses this advantage over the botanist and entomologist, that he can follow out his open air studies at all times and at all seasons. With the objects of his search there are none of those periods of appearance and disappearance such as limit his brother naturalists. Fossils are to be found in all kinds of weather, and can be studied *in situ* at Christmas just as well as at Midsummer, provided always of course, that the geologist himself be indifferent to seasonal changes.

Our visitors, however, of geological predilections are mostly here in the summer months, and never a day passes then when several may not be seen perambulating the clay and the sands, hammer and knife in hand, on a search which will undoubtedly proved successful. For East Wear Bay is a grand storehouse of the relics of bygone worlds.

> " Thick as autumnal leaves that strew the brooks
> In Vallombrosa,"

are the tokens that tell of the ancient inhabitants. Glittering in the blue clay as the tide recedes are innumerable shells of Ammonites more or less imperfect, and the valves of Inocerami in a similar condition. Perhaps a few hints on the best mode of proceeding will prove not unacceptable to those unaccustomed to working in the Gault. A hammer, generally regarded as a *sine qua non* on a geological expedition, will be useless here, there is no hard

E

rock, unless you turn your attention to the blocks of
sandstone or chalk, in which, however you would
find few fossils. A good broad-bladed knife, stout,
but flexible, is the required weapon. It is useless to
try to get out any fossil by itself, separate from its
clay bed; it will undoubtedly fall to pieces during the
process. Whenever you see the brilliant nacreous
shell, or a portion of one, jutting up above the clay,
remove as much from it on the upper surface and at
the sides as will display its shape and characters
sufficiently; then *dig* it out with a block of the clay
itself attached, on which it shall remain as if you
had intentionally mounted it, shape the block carefully with your knife (best done after you get home),
and leave a thickness of half or three quarters of an
inch according to the size of your fossil. This will
gradually harden, and in time will be like a little
block of stone. The shell itself and likewise the clay
will be all the better for two or three coatings of thin
gum put on at intervals. If these directions are
followed out I think you will find your specimens
last for years, and preserve much of their beauty.
But once again, do not try to get them detached
from their matrix. Of course you will find plenty of
detached specimens rolling about in the bay, and
probably a few good ones, but there Nature herself
has performed the work for you—take advantage
of it.

Neither will it be of any use to work in the dry
clay above high water mark, sorely tempted as you
may be to do so by the sunlight reflected from a
dozen glittering fragments of embedded shells. No
specimens can be extracted therefrom otherwise than

by instalments. Under your hand the gault divides itself into little cubical blocks, and the shells are all disintegrated. Follow the tide as it goes out, and *work in the wet clay.*

The section exposed from the Harbour at East Cliff, and stretching eastward towards Dover, is a good one, and highly instructive to the young geologist. I have mentioned the general nature of the cliffs in the Lower Sandgate Road in a former ramble. These cliffs are but a continuation of them, Old Folkestone standing in a valley formed by some river cutting through them in ancient times. They belong to the *Folkestone Beds* the upper division of the *Lower Greensand*, or, as some prefer to call them the *Neocomian Beds.* A few fossil oysters and terebratulæ, with an occasional pecten, &c., may with difficulty be extracted from them. They dip rapidly here into the sea, you may follow the rocky strata down to low-water mark.

Above them comes the *Gault*, but between the two a curious bed known as the *Junction Bed,* about eighteen inches thick, variegated with yellow, red and black, and glittering with crystals of selenite and iron pyrites; full, as full as it can be, of phosphatic nodules with which are mixed ammonites and black wood bored by ancient worms. Large lumps of selenite may be found among the rocks at the foot of the cliffs. What is the origin of all these nodules so highly charged with phosphate of lime? Since, so far as we know, this substance is mostly an animal production in Nature, they would appear to tell of the decayed remains of countless generations of creatures which had lived in the old Gault seas.

Thickly strewn about and water-worn as they are all through this seam, they must represent such remains which gradually accumulated on an old sea bed, tolerably clear of sediment. Inland this bed appears to thicken considerably, and was worked a few years ago at Cheriton for the extraction of the phosphate for agricultural purposes. It forms here a good line of demarcation between the Greensand and the Gault. In the Gault itself, however, there are two or three similar beds to be traced.

The *Gault Clay* here is about 100 feet in thickness, and has been divided by Mr. Price into eleven beds characterised by particular species of ammonites.

Between the Gault and the Chalk we have a feeble development of the Chloritic Series or *Upper Greensand*, blocks of which, some I believe still *in situ*, may be seen on the floor when the tide is out. Fossil wood may occasionally be found in it.

Then come the *Lower Chalk* strata with an almost entire absence of flints, and a scarcity of fossils. But between the high chalk cliffs and the Lower Greensand, *i. e.* over nearly the whole space occupied by the Gault, the ground here is all in confusion

"Crags, knolls, and mounds confusedly hurled,
  The fragments of an earlier world,"

and confusion gets worse confounded every winter (see p. 29). So that it is difficult, even impossible to trace the order of succession anywhere except on the bed of the sea itself.

Among the most striking objects on this ancient floor once again laid bare, are the lumps of iron pyrites—sulphide of iron, strewn about in abundance.

Very pretty specimens may occasionally be picked up in the crystalline form, but all are interesting from the curious shapes they assume. Many of the fossil shells have become completely impregnated with this substance so as to become veritable iron shells. Veins of it too, may be traced here and there in the clay.

It would not be possible to give here a list of the many hundreds of species of fossils that have been found in the blue clay but I may mention some of commoner kinds. A skeleton, almost entire, of one of the ancient Mesozoic reptiles was discovered at Copt Point very recently by Mr. Griffiths; it was probably over twelve feet long, and has been named *Mauisaurus Gardneri*. A quantity of smooth quartz pebbles was found with the bones, which, it is probable, had been swallowed by the animal, as pebbles of all kinds are absent from the Gault. Teeth and vertebræ of fishes of the shark tribe are to be found; and the remains of crabs and lobsters are far from uncommon.

The most attractive fossils in the bay are the Ammonites and Belemnites. Of the former may be found :—

A. splendens.	A. lautus.
A. auritus.	A. tuberculatus.
A. varicosus.	A. interruptus.

and many others. The commonest of the Belemnites are B. *minima*, and B. *Listeri*. Of other chambered shells we get *Hamites, Scaphites, Ancyloceras, Nautilus*, &c.

Of univalves we get the following genera :—

Rostellaria.	Natica.
Turrilites.	Trochus.

Scalaria.	Turbo.
Pleurotomaria.	Dentalium.
Solarium.	Bellerophon. &c.

Of bivalves :—

Inoceramus sulcatus.	Plicatula.
I. concentricus.	Pecten.
Nucula ovata.	Lima.
N. pectinata.	Terebratula.
N. bivirgata.	Rhynchonella.
N. ornatissima.	Arca. &c.

Subjoined is a table of the formations of the neighbourhood in the order of their succession, the newest being at the top :—

BRICK-EARTH, in scattered deposits, containing bones of the Mammoth and Rhinoceros.

PLEISTOCENE DEPOSIT, in the cliff at the back of the Pavilion, containing freshwater shells, with bones of Mammoth, Hippotamus, Rhinoceros, Hyæna, Irish Elk. Oxen, Horses, &c.

UPPER CRETACEOUS.

UPPER CHALK with flints, (near Dover).

LOWER CHALK with few or none.

CHALK MARL.

UPPER GREENSAND.

GAULT.

LOWER CRETACEOUS OR NEOCOMIAN.

FOLKESTONE BEDS.

SANDGATE BEDS.

HYTHE BEDS, (Kentish Rag).

ATHERFIELD CLAY, (beyond Hythe).

OCCASIONAL PAPERS.

———:o:———

I.—ON THE STUDY OF NATURAL HISTORY.

Read before the High Wycombe Natural History Society, April 1866.

———

" GOD fulfils Himself in many ways."

The study of Natural History may be looked at from two points of view; we may regard it either as affording pleasure to the senses and gratification to the mind; or as tending to be practically useful in the economy of our lives. It is now closely followed up by the holders of each of these views, and no student in either class has ever repented the study. Nature herself is so infinite and varied in all her productions, that though she has had disciples ever since man appeared on the earth, she retains, even now, after the lapse of thousands of years, the same freshening influence, the same charm hanging about her works, which acts with such an irresistible force upon the neophyte and urges him to travel onwards. It is not my intention now to refer at all to the advantages derived from the study by those holding the second view; we are assembled here as we have been at other times, simply from a love of Nature, with a desire so to look upon created works, that we may find "life and food for future years." To many I may say nothing new; to some I may probably be able to place some old facts or thoughts in a new light; but I shall be amply repaid if I succeed in making only one more eager in his or her pursuits in the woods and fields—more desirous of following out thoroughly that which at present is taken up only in a desultory manner.

I believe the love of created works to be inherent in the human mind—that it is not so much an acquired love as one that will spring up involuntarily; we have it in us naturally; it may lie a long time dormant, but when some flower of spring, or animated "thing of beauty" shall appear, at a moment perhaps when the heart is peculiarly open to its influence, it will implant itself in our memories, and become a "joy for ever." Few indeed are they, who, having once set foot within the porches of the great palace of nature turn round and retrace their steps. And the farther they advance the greater is their wonder and delight—the more keen is their sense of enjoyment. When LINNÆUS, after years of study, came to England, and for the first time in his life saw the yellow gorse in flower, he fell on his knees, and thanked GOD for the sight. No one can understand this who has not discovered a rare plant or seen some beautiful animal for the first time, that he has long wished to find.

Just as in childhood, as the years—nay, as the weeks—roll by, we make fresh discoveries in the world around us, feel ourselves growing wiser—feel an expansive power at work within us, produced by the very objects which that power enables us to appreciate—so do we, in maturer years, among the domains of nature, feel sources of new pleasures

ever opening to us, and we make continually new
discoveries. The things which delighted us in childhood, yield us little delight in manhood—*then*

> " Earth, and every common sight,
>   To us did seem
> Apparelled in celestial light,
> The glory and the freshness of a dream."

But a sort of wearisome familiarity began to cling
to them,

> " Shades of the prison-house begin to close
>   Upon the growing *boy*,
> But he beholds the light, and whence it flows,
>   He sees it in his joy ;
> The *youth* who daily from the East
> Must travel, still is nature's priest,
>   And by the vision splendid
>   Is on his way attended ;
> At length the *man* perceives it die away,
> And fade into the light of common day."

So, says the poet, is it with the ordinary experiences
of life.   If it could be shown then, that there was
any one subject of study, which, beyond all others,
and with less trouble, could afford us a never-ending
experience of *new* pleasures—pleasures, which
should not pall our satiated appetites, which have
the very least alloy of disappointment in them, is it
not worth while to pay a little attention to it ?   I
may be said to be exaggerating, to be enthusiastic in
my mode of recreation ; but I appeal to all naturalists to bear me out in what I have said, and I
confidently leave it to the experience of others.

The subject is one, not so much for the library and
the study, as for the theatre of Creation itself—you
will bear in mind the view with which I am now regarding it—we shall learn most by personal examination, and what we so learn we shall seldom forget.

Nature probably is most fascinating, subjectively,

in the season of youth, the mind being then most
capable of pure enjoyment for its own sake ; all
things then wear a fairy garb ; it was then, says
Wordsworth, that

> " The sounding cataract
> Haunted me like a passion : the tall rock,
> The mountain, and the deep and gloomy wood,
> Their colours and their forms, were then to me
> An appetite ; a feeling and a love
> That had no need of a remoter charm
> By thought supplied, or any interest
> Unborrowed from the eye."

And as riper years steal upon us the same love retains its hold, but there is a change in the mode of
regarding it ; we, like the poet, learn

> " To look on Nature, not as in the hour
> Of thoughtless youth, but hearing oftentimes
> The still sad music of humanity,
> Not harsh nor grating, though of ample power
> To chasten and subdue. And we have felt
> A presence that disturbs us with the joy
> Of elevated thoughts ; a sense sublime
> Of something far more deeply interfused,
> Whose dwelling is the light of setting suns,
> And the round ocean, and the living air,
> And the blue sky, and in the mind of man :
> A motion and a spirit that impels
> All thinking things, all objects of all thought,
> And rolls through all things."

To come to something practical : let us draw a
comparison between a lover of nature and one who
thinks nothing of her. Take the case of a simple
ramble through the fields : most people are in the
habit of " doing a constitutional " occasionally.
This walk is very often quite aimless, and is only
undertaken as a matter of duty, out of regard to
one's health. A man takes a certain number of
steps every day ; he feels a sort of satisfaction after
it, and goes to his work again until the time returns

for its repetition. All well and good perhaps, but I
ask, is it not also our duty to keep our minds in
health, as well as our bodies ? The above individual
grows no richer, mentally, for his labour. How
different from the case of another, who tells you he
never comes home from a ramble without having
discovered something fresh : he goes out to escape
from his daily routine of business ; he knows that
nothing rests the mind so much as *change*, and that
when it is thoroughly wearied out by continued con-
centration on one subject, it is better to occupy
it with another than to suffer it to be idle. And
therefore in his walk he notices the flower and the
animal, their habitats, and their times of appearing ;
he discovers, without the aid of books, that there is
" a time for everything "—a set time, and that in the
beautiful regularity which pervades nature, nothing
appears out of time or order ; the caterpillar is not
hatched before its food-plant is putting forth its
leaves ; the butterfly and the bat do not wake from
their winter's sleep when there is nothing for them
to eat ; everything is arranged. He notices, with
scarcely an effort, the peculiarities of the beasts of
the field, and the birds of the air ; he discovers the
marvellous connection between one species and
another, between one family and another, and the
dependence of all upon the Creator, so that

> " The whole round earth is every way
> Bound by gold chains about the feet of God."

In the Spring *his* eyes first see the swallow, *his* ears
are first greeted by the cuckoo, he is gratified by the
bursting forth of the vegetation into the most lovely
green ; in the Autumn, while tints still more lovely

objectively array themselves before him, his delight
is tempered with sober thoughts of the great change
which is one day to be wrought in himself. In Summer
he beholds the triumphant reign of all living things,
and in Winter—generally thought to be dull and
cheerless in the country,—he knows where to find
the squirrel and the dormouse snugly domiciled;
he can find you the chrysalis of many a moth and
butterfly marvellously entombed in the earth, or
slung in a hammock; he can show you luxuriant
beds of mosses—those children of the winter that
flourish when all around is asleep. And even if
he could not *show* you all this, think what
marvellous stores of information he has laid up,
that shall afford him food for thought when he is
lonely, or from which he can draw fairy lore to
while away the winter evening; what tales he can
tell you of the wonderful things he saw in the
summer—how he found the boat of eggs floating
about in the pond, so curiously and perfectly formed
by the gnat, that it could not be upset—a veritable
life-boat; again, how he drew from the water a
thing monstrously strange, armed with jaws that
could unfold themselves upon its prey while yet afar
off, how with unrelenting stedfastness it destroyed
and devoured the other inhabitants, and after a few
months of such enjoyment it climbed up a tall reed,
and splitting itself down the back, took unto itself
wings and flew off to continue its carnage among the
inhabitants of air. Or our naturalist may give you
more pleasing accounts of the nests of the wren and
titmouse, the beautiful spotted eggs of the thrush,
and the pearly eggs of the azure halcyon—how one

bird assailed him with a torrent of abuse as he
approached her offspring, and another suffered him
to lay hold of her, sooner than she would forsake
her nest: again, of the banks of flowers upon
which he lay and pondered—the bed of happy
violets, the golden cowslips, the "jocund company"
of daffodils, the delicate wood sorrel, and the wind
flower; he tells you how he saw the face of wintry
nature turned into a perfect paradise of loveliness,
and says—

> "Though absent long
> These forms of beauty have not been to me
> As is a landscape to a blind man's eye;
> But oft, in lonely rooms, and 'mid the din
> Of towns and cities, I have owed to them,
> In hours of weariness, sensations sweet."

These are the stores upon which the lover of nature
can draw.

The poets of nature have been many, and I must
not take up your time in quoting what is most likely
familiar to you. I have tried to show what a charm
there is around us if we like to experience it—what
an infinite variety there is for the mind to study.
It is this infinite variety which gives the superiority
to Natural History as a means of recreation: there
is no fear of exhausting the subject. Alexander the
Great was sorely distressed when he had conquered
all there was to conquer; but it cannot be so with
us, Creation knows no limit. I remember in some
"wild dream of a German poet" that a human
being was conducted over the universe to view God's
worlds, and that after sweeping past innumerable
orbs,—planets, satellites, and comets, the mind of
the man sank into itself, and shuddered with the

over-powering effects, begging to be shown no more. If it were so with the thought of the infinity of *worlds*, what would it be, could he have but a dim comprehension of the infinitude of infinities that exists in each separate world.

Here then is provided for our delectation a goodly storehouse of knowledge ; volumes on volumes lie open before us ; take them up and reverently turn over the leaves, they make up the Book of GOD.

II.—WINTER WORK.

——:o:——

Read before the Folkestone Natural History Society, December 2nd, 1868.

——:o:———

The glorious summer weather of 1868 is all past, and the usual October and November gales sent us rather sooner than we expected into the regions of winter. All around us now is inhospitable and bleak, and there is little inducement to follow out in the open air the practical study of Natural History. So we are tempted, perhaps, to sit still and ponder over the rambles we took in the summer, to regret that they are over, and to wish they would soon come again. It is well, perhaps, that we should do so, for they ought to have supplied us with a whole treasure-house full of " studies," from which we may draw one after another to gaze and admire. It is well to ask ourselves now, with these pictures set in the golden frame of memory still fresh before us, whether we valued them so much before they were thus framed—in plainer words, whether we thought at the time that we really had great opportunities for gathering food for thought in quiet hours. Did we do all we might have done ? In what respects did we fall short ? So we may gather experience to guide us when the swallow and the cuckoo return again. Perhaps some of us made a tolerable collection of objects, which we had not then time to arrange, perhaps not even to name. Now is the time; the collector would never get through his work if he were always collecting, if he never had seasons of leisure, for simply *gathering* objects is but half the work; they have to be compared, classified, and specified; general laws deducted, hasty conclusions tested, perhaps reversed; all this you know is specially the work of the mind, and the mind at such times must not be hampered by the body, leisure and freedom from disturbance is essential for contemplation and study. There are our Land and Water Shells; it was no easy task to name some of them, and no doubt some of us have got three or four pill boxes full of shells somewhere or other labelled " doubtful." Now is the time, when the drizzling November rain keeps us in doors, to sit down, and by the aid of Lovell Reeve and a magnifier to settle the question. A few papers of dried flowers too, not yet properly labelled, will occupy us now and then for an hour or two, perhaps also birds' eggs, seaweeds and fossils. Winter is necessarily the time for theoretical study; we cannot do so much out of doors, and in Summer, when all is favourable for so doing, it would be folly to be reading books at home. We must perfect and complete in December what we began in the early spring.

But there is an impression I know that no active out-door work can be done by a naturalist in winter. I should be glad to do something towards removing this impression. Winter is not so lifeless as we are apt to think. I look back to many mild days in December and January when I experienced great pleasure and gained no little knowledge in my rambles—days spent in the leafless woods perhaps, but yet where the squirrel might be surprised at a winter meal, and the hawk at its feast of blood. True, there is not in winter the mysterious abundance of life around us which astonishes us in summer, but the very lack of this abundance renders it easier for us to make observations on those objects that are left. In June and July we are so embarassed by the multitude of objects we see around us, that we do not know where to begin, we feel quite helpless till some friendly hand comes and puts us to work. Now Botany is a subject which is associated so thoroughly with summer that few ever think it possible to do anything at it in the cold weather. Yet winter has a flora of its own, and even now, in December we might go and gather a handful of flowers. There is more room however for active work among the mosses which flourish most luxuriantly in the midst of snow and rain. Many of them ripen their fruits only in the dead of winter, and for beauty of detail, they rival all the rest of the botanical creation. It is worth a damp walk to some woody dell to see their varied hues of green, and the marvellously contrived mechanism for the dispersion of the fruit which characterises them; they appear all the more flourishing by reason of those very influences which lay their more sturdy brethren low. And many an evening's amusement may be obtained by studying these mosses with a microscope.

In Geology a great deal can be done; when neither plants nor animals put in an appearance we

F

can always go geologising. I do not mean simply
fossil-hunting, but geologising in the fuller sense of
the word—gaining a knowledge of all the formations
in our neighbourhood, where they crop up, and their
line of strike—the gravels, the clays, the sands, and
the drift, as well as the harder rocks; all these will
afford plenty of room for speculation, too much,
perhaps, but at any rate they will set us a thinking.
And if we go out simply in search of fossils, we
shall meet plenty to encourage us in this rich neighbourhood. Among the chalk on the cliffs crowded
with fragments of *Inocerami* and *Rhynchonellæ*; in
the Greensand blocks scattered over the beach in
East Wear Bay, rich in oysters and fossil woods;
and above all in the blue clay left bare by the
receding tide, studded with countless ammonites
and belemnites—here we shall find ourselves surrounded by the remains of a former world, and find
problems set us that men very far wiser than
ourselves have never yet been able to work out.
Although it is certainly not pleasant to stand
chipping corners off stones with a cold hammer,
with the wind and sleet driving in our faces, yet
there are many mild soft days in the very depth
of winter when we may thus comfortably amuse
ourselves.

But now, to pass to the animal world. Many of
our sylvan inhabitants have, it is true, retired for
the winter, but they are not wholly lost to us. Many
a time, when rooting among the mosses we shall
turn out perhaps a beetle snugly ensconced, or a
plump caterpillar, perhaps a dormouse fast asleep,
and this sets our mind astir in another direction, we

ponder over the mystery of *hybernation*. *Why* these animals should thus pass away the winter we can perhaps see—change of climate and scarcity of food render it necessary ; but *how* it is done is beyond our ken. We see it in all sorts of creatures—in the great Brown Bear in the forests of Russia, in the Marmot, in the Squirrel, down to the tiny caterpillar not the tenth part of an inch in length. We see it again in those weird creatures the Bats. Go into some sheltered cave and you will probably find numbers of these creatures hanging by their claws to the roof, head downwards, their wings closely enwrapped around them—not a sign of life, not even any perceptible breathing. It is not merely sleep, you may rouse up an animal from its ordinary sleep, and it does not take long to collect its faculties ; unlike the lords of the creation, there is no stretching of limbs and rubbing of eyes, the creature springs up from slumber and is on the alert at once. But not so with hybernation ; it takes some time to rouse a bat, the wakening comes very gradually and is generally fatal. It is evidently a much nearer approach to death than sleep is—the breathing is so slight as to defy investigation, and the blood courses so sluggishly along that you can detect no pulsation ; the air in which the creature passes the winter, undergoes no change, and strangest of all the animal will exist for some time in gases that would be immediately fatal to it if awake. I just refer briefly to these points in the hope of provoking a discussion on the subject presently, By thus exploring caves and other suitable spots, we may become acquainted with some species of bats not otherwise often seen. I remember once going into a chalk cave and finding four species, the Pipistrelle, Noctule, Long-eared Bat, and the Lesser Horse-Shoe Bat having the curious leaf-like appendage to the nose.

Again we may study *birds* as well during the winter as in summer, perhaps some species better. The little Tits may be seen in flocks of about a dozen flitting in and out of the hedgerows, or busily running up and down the stems of trees searching for sleeping insects; the little Wren often scuds across the road a foot or two above the ground ; the song of the Skylark may be heard on a sunny day in any month of the year. The habits of most of our birds change as they don their winter plumage ; they begin to flock together in great numbers, especially the Starlings, the Larks, the Finches, &c. The Chaffinch is seen in large flocks, containing only males, very few females are to be seen, and these mostly in the south. Another question here arises—Why this collecting together in flocks ? And why in winter and not in summer ? Well we, perhaps, can understand *why not* in summer, because of the family duties which engage them, and the intense rivalry and jealousy of the males. These feelings, however, die away with the summer months. Do they congregate in the winter for warmth, or for food ? Scarcely the latter, since it would be easier to obtain food singly.

Then there are the birds of passage—those going and those coming; the Swift, the Swallow, the Cuckoo, and others disappearing; the Fieldfare, the Redwing, the Hooded Crow, &c., coming. The latter, as of course you know, frequents the shore and the adjacent fields now in search of food, and at once attracts notice by the hood it wears. Where does it raise its family? does it ever breed in England? Why does it come here at all?

Migration is almost as wonderful as hybernation. Before it was so well established a fact as it is in the present day, hybernation was much more extensively allowed. The Swallow tribe in particular attracted most notice, as was but natural, and they were all firmly believed to spend the winter in this country, hidden up in caves and rock crevices, old buildings and places similar to those where we find the bats; some thought even in the bottom of lakes and rivers buried in the mud. Dr. Johnson in his usual dogmatic style, once remarked in the course of conversation—" Swallows certainly sleep all the winter, a number of them conglobulate together by flying round and round, and then all in a heap throw themselves under water and lie in the bed of a river." And Gilbert White of Selborne, could never bring himself to totally disbelieve in their hybernation. Nor has the belief died out in the present day, for there was a discussion about it in the pages of " Science Gossip " only a few months ago. But this is rather digressing.

There is often much talk about the " mysterious " instinct which guides birds in their migrations. I confess I can see little mystery in it, not nearly so much as in hybernation.

Disbelieving totally, as I do, in what is commonly called the " instinct" of the lower animals, and believing that the whole animal creation possesses pretty well the same faculties and reasoning powers as ourselves, nay, I may go further and say, an immaterial and undying principle similar to our own, the mystery commonly supposed to be connected with the instinct of animals, vanishes in my mind to a considerable extent, It is improbable in the highest degree, that a flock of birds all of the first year, should set off for a foreign land alone, with no old ones in their company, who have been the road before ; and therefore, I believe there are always plenty in a flock to guide them. And if so, why should not birds be able to travel about just the same as men ? But even supposing for the sake of argument that such an improbability as I have stated takes place, what then ? A flock of birds feel the weather in their locality, getting too cold for them. They do what a tribe of men might do, try to find a warmer place. If they fly northwards, they only experience colder winds, what should they do then but turn round to the south ? In that direction they meet with warmer air, and are beckoned continually on and on by more balmy breezes, until they arrive in a locality which suits them, and there they wait until they feel compelled by circumstances to go back again. In short they act like *reasonable* beings as they are. I should be glad if some one would take up the discussion of the subject presently.

I fear I have trespassed somewhat too much on your attention, and must now draw to a close my desultory remarks. I have simply tried to show what we may all do in what are generally called the dreary months of winter, and I hope I have proved that there is plenty of occupation both for mind and body.

III.—THE BLOOD BEETLE.

Timarcha lævigata.

Reprinted from *Science Gossip* for February, 1867.

——:o:——

It was in the month of October, several years ago, that I first became acquainted with the Blood Beetle. It was crawling over some herbage at a very sluggish pace, totally different to the hurrying race of a Sun-Beetle across your path, or a Weevil over the leaves, and I took it up to examine it. While turning it over, I found my fingers were covered with what I at first took to be blood; recollecting, however, that none of the other beetles with which I was acquainted afforded the sanguineous fluid, I looked a little closer, and discovered a rich scarlet bead, very translucent in appearance, emerging from the creature's mouth. Upon taking up several others they behaved in the same way, and the habit appeared evidently a defensive one, although the fluid was to me perfectly harmless; it might not be so, however, to the ordinary enemies of the Beetle. This habit, together with the firm ovate appearance, and the worse than snail's pace, at which it crawled along, made the insect very interesting to a neophyte in Natural History, and not knowing its name, I called it *pro tem.* the Blood-Beetle, which, perhaps, is slightly more refined than its common English cognomen, " The Bloody-nosed Beetle." As it was then rather

late in the year, there was not much opportunity for
discovering many of its peculiarities ; it soon retired
from observation, probably burying itself among
thick moss or herbage.   Early in January, however, it was abroad in the sunshine under the
hedges, and my attention was drawn to it.   In April
I noticed another curious creature, very common : it
appeared to be some kind of larva.   It was about ten
lines in length, of a dull metallic green above and
pinkish beneath, the whole body very wrinkled, and
in general appearance convex.   It was feeding on
Bed-straw, and, where one specimen was seen,
plenty of others were sure to be found.   It was not
until I had taken up several to look at that some of
the well-known fluid appeared, and the thought at
once struck me that the creature was the larva of
my new friend,—the Beetle.   It fed, too, on the
same food, *Galium Aparine*, and more rarely on *G.
mollugo*.   I at once collected the larvæ and caged
them, and after a time found my suspicions correct,
for they produced some very fine imagos.

This was one of my first entomological discoveries,
and, like every other beginner, I felt a good deal of
satisfaction at having made it myself without the aid
of a book.   I mention this simply as an illustration
of the pleasure awaiting anyone who chooses to
search for it in the insect world.   A few of the notes
I have since made on the same species may, perhaps,
prove interesting to some of our readers.

The Beetle itself is the most plentiful of the larger
coleoptera in our neighbourhood, being found on
every bank, and under every hedge ; it appears also
to be the most hardy, for there is probably no season

of the year when it may not be seen. I have caught it in every month except December. The larvæ are to be found in April and May on the Bed-straws, looking when young merely like small black protuberances on the leaves. At first sight it would appear that they do not possess the usual number of prolegs or claspers—so prominent among the lepidopterous caterpillars—having apparently only one at the tail. Although Westwood mentions this as single, it is evident to the naked eye, and much more so through a glass, that it is a double one, quite as much as that of a hawk-moth larva; the other four pairs are present in the shape of small tubercles on the abdomen, and are seen quite plainly if the creature be allowed to cross the hand held up horizontally to the light; each is then seen to be brought into full play in the act of walking—they are not so easily detected when it is crawling over the herbage. When seized it rolls itself up like a hedgehog, not being proportionally long enough to do so after the fashion of larger caterpillars. When alarmed, I have known it, in various instances, to emit the scarlet fluid, but it is not done so freely as by the imago. It changes its skin at regular intervals, appearing immediately after of a reddish hue, particularly about the head and legs: it gradually darkens in colour. The larva is quite as sluggish in its movements as the perfect insect. All my specimens were buried by June 10th, but some had gone down into the earth a fortnight before. On July 4th I disinterred one or two; they were then of a very light pink colour, very jelly-like in appearance; the legs were perfectly formed, and the wings lay loosely by the side of the body, which was on its back. A small cavity had been formed in the soil in the usual way, the sides of which were made quite compact by the pressure of the body, and at one end lay the cast-off skin, the antennae were full size, but these, like the legs, were pink, much deeper in shade than the body. On July 22nd I disinterred another, almost perfect, but the body was still soft and pink, while the elytra were of their proper hue : a second specimen was still without the wing cases. The first imago emerged July 30th, and this was soon followed by others.

The perfect insect is very ovate in appearance and firm in consistency, the under surface and also the legs are of a glistening metallic dark blue, the elytra are nearly black, as is also the head and thorax. It is placed in the family CHRYSOMELIDÆ, and in the genus *Timarcha*, though formerly it was called a *Tenebrio* : the specific name is *tenebricosa*, or more lately *lævigata* : the latter term is preferable as it serves more directly to contrast it with the other species in the same genus—*coriaria*. The elytra are soldered together longitudinally ; and when they are forced open the Beetle is found to be wingless ; it is thus totally incapable of flight, and is the largest vegetable feeding insect in England so constituted. The tarsi are very broad, and afford it the power of taking a firm hold of the herbage over which it crawls. At night it rests clinging to stems with its head downwards ; they are difficult to discover in the early morning, being covered with heavy dew, and looking more like dry seeds than anything else ; as soon as the heat of the sun has caused all the moisture to evaporate, they begin their peregrinations. The antennæ look like strings of small beads, very beautiful; and, when the insect is moving, they are in constant motion from side to side, tapping the ground or stem over which it is travelling, as if to test its safety. The scarlet fluid is said by Westwood to be emitted both from the mouth and the joints of the limbs; I have never, however, been able to detect the smallest particle flowing from the latter places. Country people say it is a specific for the toothache; and, having once tried it, I am inclined to believe them. I found relief from rubbing it over the tooth and gums; but, perhaps, one is not entitled to state it as a general fact from one trial. The Blood-Beetle is often figured and spoken of in old works as the Catch-weed Beetle, no doubt from its being commonly found in *Galium Aparine*.

IV.—ABOUT THE SLOW-WORM.

Reprinted from the *Naturalists's Circular* for June, 1868

——:o:——

I remember one of my companions telling me how during a ramble after wild-flowers a slow-worm had sprung at him from the bank, and only just missed fixing itself on his face. As he firmly believed the creature to be of a deadly poisonous nature, he regarded this as a wonderful escape. Neither of us knew much of natural history then, and I accordingly congratulated him on the fact that he was still alive ; indeed, to tell the truth I felt rather envious, and wished that I could relate such a marvellous escape. But this was many many years ago, and the slow-worm has since then been one of my numerous pets. A box of these reptiles is certainly a novelty to most people, for though they may have come across one or two in their occasional rambles, it was only to jump away from them, or to strike at them with a stick ; whereas before a box with a glass front they can gratify their curiosity, and at the same time feel perfectly safe. As you dilate on the curious traits of your little protegés to an admiring audience, they listen to you with about the same feelings as those experienced by the crowd before a cage of tigers while Mattoko the great tiger-man is giving a lecture on his captives. Will they test the creature's harmlessness by taking one into their hands? " Ugh!" (quite involuntarily) "no, *thank* you!" I recollect taking a few once to a natural history meeting (where certainly folks ought to have known better) and one happened to get on the floor. It was ludicrous to see how instantaneously that floor was vacated, and every lady was standing on a chair. Poor *Anguis fragilis*! he was picked up by a friend and replaced in his box, much to his own satisfaction and that of the ladies.

Slow-worms used to be very plentiful in the Quakers' burial ground in a town where I formerly lived, and were killed by dozens when the grass was mown, the man using a stick about ten feet long, so that he might be out of danger. On lifting up the flat stones we often came upon six or seven at once, and although the creature in a general way deserves its name, it can in times of peril make off pretty briskly as it attempted to do then. On being seized it twists itself in and out between one's fingers in a manner peculiarly unpleasant to those not accustomed to it. I took home three or four and put them into a good sized box with some earth and rockwork, beneath which they soon formed regular hiding places. They are not all of the same colour, some being of a rich sienna brown with darker markings, and others of a leaden hue; whether this denotes distinction of sex, or is a mark of age, I am not prepared to say for certain, but all the old ones I have ever seen were of the dull hue, and I never saw any young ones that were not bright brown. I came across one at six a.m. in a ramble up a lane. I was hunting for molluscs and heard my dog barking as if she had made a discovery. Such I found was the case, she had surprised a slow-worm as it was retreating, and without endeavouring to touch it was performing a circular dance round the hole it was entering. I seized it before it made good its escape, and carried it home to the others.

The general food of the Slow-worm consists of the small white and grey slugs so common and so mischievous in gardens, and the mode of eating is very peculiar. If hungry, on seeing a slug crawling along it approaches it, eyeing it intently, and with the greatest deliberation seizes it across the middle ; there is no darting, nothing sudden,—it merely opens its mouth and quietly takes hold of its prey. Of course, from the nature of its food, we see there is not the slightest necessity for rapid movements such as we find in those of its near relation, the Scaly Lizard which has to catch flies. So with the swallowing, it is done very gradually, and often takes a long time, a considerable quantity of fluid covering the mouth meanwhile. If the slug has been crawling over the earth, and has anything adhering to it, the slow-worm will take it to a stone and rub it against it till it is detached. This shows the the possession of a considerable amount of reasoning power, as it was only done when necessary. I could not always provide slugs for them and sometimes they took earthworms, but they did not relish them so well.

Like all the members of its class the Slow-worm is under the necessity of changing its skin at intervals ; it does not, as with the snake, come off entire, but in several detached portions, which peel off, the creature assisting in the process by twining its body in and out amongst rough substances. One of mine changed on May 1st, and again July 14th. The common name of "Blind-worm" is so utterly inappropriate that I cannot conceive how it arose, and yet many country people will tell you that it cannot see at all. But it is very evident that it discovers its prey more by sight than by any other sense, and its eyes are peculiarly pretty and gentle, without the baleful aspect of those of the viper. My slow-worms buried themselves in the garden during the winter, and the first reappeared March 28th, very sluggish and sleepy, and it ate nothing for several days. The property of throwing off its tail and reproducing it must be well known to my readers.

On the Warren here they are very plentiful under the large stones. Their usual length is about thirteen or fourteen inches, but I had one brought which measured nineteen and a half, of which the tail took up eleven ; this was a monster.

I began with an anecdote, and I will end with one. A friend of mine was hunting for slow-worms underneath some stones, and wanted very much to move one immense rock, having an intuitive knowledge of the presence there of a host of victims. But it was beyond his strength, and he called in the assistance of the military in the shape of two soldiers who happened to be passing by. When the stone had been overturned there lay several slow-worms, among which my friend immediately darted, seizing two or three in each hand. The sight was too much for the soldiers who fled precipitately, waiting for neither thanks nor largess.

V.—THE LARGE GREEN GRASSHOPPER.

Acrida viridissima.

Reprinted from *Science Gossip*, September, 1868.

——:o:——

Among the terrestrial creatures now peopling the earth in such vast myriads, none come across our path so often as the Grass-hoppers. Every tuft of grass seems alive with their curious calls; at every step one seems to hop away from us. And they well repay a short examination, were it only for their varied and lovely colours : some having their habitation solely in the grass, are of its own green hue, others enlivening the bare chalky slopes are grey while some again are of a delicate carmine, some green and some crimson—in fact, as we catch one after another, there seems no end to their tints. The Large Grasshopper, of which I have made a few notes, is, however, much more seldom seen ; when it is caught it is generally killed and pinned out, as something a little out of the common, but very rarely is it kept alive, and consequently little or nothing is known of its habits. My friend Mr. Tate just mentions it in the volume for 1866, and Mrs. Watney also states one fact concerning its food. Were it better known I suppose we should have a recognized English name for it ; as it is, it goes by the name of the Green Locust, or the one at the head of this article. But, scientifically, it is neither a locust nor a grasshopper, both of which are included in the family *Locustidæ*, but this belongs to the *Gryllidæ* which is distinguished from the former by the presence in the females of a formidable looking ovipositor extending from the end of the body. Mr. Tate calls it the " Horsehead " Grasshopper, but I do not see why the title should be applied solely to this species, as all of the tribe have heads much alike ; perhaps the resemblance to a horse's head is more striking in *viridissima* from its great size.

I caught a female two or three weeks ago as it sat on the head of a large flower, and brought it home captive. It has been living in a glass globe ever since, and at the present moment is doing a constitutional over my writing table ; not, I fear, appreciating the honour I am doing it by giving its history in the pages of *Science Gossip ;* at any rate, it does not look as if it did, being busily engaged in discussing the contents of a cabbage stalk I purposely laid in its way. I did not know at first what food to supply it with, animal or vegetable, until I thought of Mrs. Watney's note above referred to. I put in some grass well moistened, a piece of cooked beef, and, as I knew the latter would not easily be obtained in the creature's own haunts, a couple of houseflies as well. It treated both flies and grass with great unconcern ; but when one of the *antennæ* came in contact with the meat, it went up to it at once and devoured it, holding it between its two front legs. Shortly after I found that one of the flies was gone, and, as it could not have escaped, I concluded the creature had eaten it, so I sat down to watch the fate of the other. *Viridissima* had retired to the

G

gauze over its dwelling, hanging with its back downwards and appeared busily engaged in " cleaning its
teeth ; but presently the fly came between its legs
and the roof, there was a sharp, sudden movement
of the head, and the prey was entangled in the complicated jaws of the grasshopper, by the agency of
which it was speedily tucked in and devoured.
Having once heard of an individual of this species
devouring its own leg which had been accidentally
knocked off, I put in a few small grasshoppers, to
see whether they formed part of its diet.   It was not
long before one of them ventured within range, and
escaped with the loss of a leg.   A few minutes afterwards another was caught bodily, and eaten with
great relish.   I now fed it entirely with living prey,
and though I put in both raw and cooked meat, it
never again touched either.   I also put in a worm,
thinking that probably, like a mole cricket I once
kept, it would eat it, but it did not.   I should think
very probably it would do so, though, if hungry.   It
is not, however, wholly carnivorous, for, as before
mentioned, it is fond of the succulent stalks of cabbage, and possibly of other similar substances.

It never appeared to hunt its prey, never ran after
the flies or sprang on them ; it waited quietly until
they came within reach, when it turned its head
sharply round and seized them, the legs assisting.
It very likely obtains its usual food by lying in wait
as it certainly could not capture either flies or grass-
hoppers in fair open chase.   Its mode of progression
appears to be a series of short leaps, on an average
twelve to fifteen inches, not nearly so long as those
of its smaller relatives, though when jumping from

a table or any elevated spot to a lower eminence I
have seen it go above a yard. It walks about a
great deal but I do not think it ever flies; when I
threw it into the air it never attempted to do so, it
spread out its wings but merely to break the fall, in
fact, it always uses its wings when leaping.

I noticed the frequent application of the tarsi to
the mouth when walking; it took place not only when
the animal was climbing up the smooth glass, but in
going up a wall or door, and even in walking over the
carpet. On the wall it took place about every half-
dozen steps, but not so often on the door or carpet,
and still less frequently when walking up my coat.
I scarcely think with Mr. Tate that by this action
the feet are rendered more glutinous, but that it
brings them to its mouth simply to cleanse them
from any particles of dust that may adhere, by which
the action of the pads or suckers may be impeded;
as it uses these suckers more in climbing smooth
surfaces than in walking, the cleansing would in the
former case be necessarily more frequent. Whenever it finds one of its feet slipping about on the
wall, it brings it to its mouth and cleans it, after
which it adheres very well until clogged again with
dust. I noticed especially when it was trying to
gain a footing on a dusty ledge above the door, it
completely failed, although every foot was repeatedly
cleansed. _Viridissima_ is very clean in its habits,
washing its face with its feet very much as a cat
does; it also with its mouth frequently cleanses its
ovipositor and antennæ, the latter being bent down
to the jaws by one of the front legs and drawn up-
wards with a curve, like that of a carriage-whip. I

have been hoping to see it use its curious ovipositor and lay some eggs, but have not hitherto been gratified. The length of my specimen is two inches and three-eighths, the antennæ are two inches, and the hind legs two and a half.

VI.—ON PRIMROSES AND THEIR FERTILIZATION.

——:o:——

Read before the Folkestone Natural History Society, March 3rd, 1870.

——:o:——

" A Primrose by the river's brim,
A yellow primrose was to him,
    'Twas that and nothing more."

I have chosen this quotation simply because it is not all suitable for the object we have in hand, nor to be taken as a motto by the members of a Natural History Society. Yet it is doubtless known to you all, and may serve as a " shocking example " of the state into which some people fall, who neglect to use their senses. Much more than simple " yellow primroses " are these blossoms to us. I might enlarge upon them as the flowers of childhood, round which cluster many " sunny memories " of our early days, that might otherwise have been forgotton ; as reminders of the happy careless hours when earth was ever unfolding new treasures and filling our minds with delight at its luxuriousness of beauty.

And again I might dwell upon them as above all others the flowers of spring, bright and happy messengers, coming with clear blue skies in January and February to tell us of the wealth of form and colour that will shortly dazzle our eyes as the bleak days get fewer, and at last disappear. But this would be too sentimental; we must look deeper into

Nature than this, and derive a still greater intellectual feast from a knowledge of the mysteries of life that lie hidden in the Primrose.

We discover at once, from the venation of the leaves, that the Primrose belongs to the exogenous order of plants, these veins forming a net work all over the leaf, and not running parallel from one end to the other, like those in a lily leaf. We notice, moreover, that there is no stem here : you may express surprise at this, seeing so many blossoms springing up from the root, each one at the head of a stalk. But these stalks are *pedicels* or secondary flower stalks, like those which spring from the head of a Cowslip stalk, and if you cut through a root just below where it emerges from the ground, you will notice that all these pedicels spring from one circling line marking the real stem, which in this case, is to a certain extent *suppressed*, or rather *arrested* in its growth. Occasionally, owing to certain circumstances, this peduncle shoots up to some height, carrying the pedicels with it ; then we get the variety commonly called the Oxlip, though not the true species of that name. To this I shall refer again presently.

And now, ascending the stalk, we come to this green cup at the summit, which all the members of the botanical class will recognise as the calyx. The component parts of this calyx, *i.e.*, the *sepals*, are five in number : that or some multiple of it, as you know, is the prevailing number of organs in this division of plants, though four shares the honour to some extent. But you do not find that these sepals are separate as in the case of the Rose, nor deciduous

as with the Buttercup; they are all united, hence
the calyx is *gamosepalous*, and it encloses the seed
vessels until its contents are perfected, hence it is
persistent. You recognise the five sepals, however,
by the five points at the summit.

Inside this calyx we come to the five coloured
leaves or petals forming the *corolla*, and these likewise are united—*gamopetalous*. Concerning the
colours of these petals naturalists are divided,
though it is highly probable that many of us have
no doubts upon the subject; but if you gather numbers of blossoms from different localities, you will
find them varying from light yellow to yellowish
green, and in some parts of England they are
decidedly pink. Some call the colour green because
it turns green when dried, but that seems to me an
absurd reason; this dried cowslip is blue, but I do
not suppose anyone would talk of blue cowslips.
Inside the corolla again we come to the stamens five
in number; and here mark a peculiarity. It is
another general rule in the arrangement of the
organs of the flower, that each member of a circle of
organs is inserted opposite to the *opening*, between
two members of an outer or inner circle; e.g. each
petal is inserted between two sepals, though inside,
each stamen between two petals and so on. But
here, as you see, the stamens are inserted *upon* the
petals (hence styled *epipetalous*) and opposite to each
—not between them; in a case like this it is considered that an outer circle of stamens has been
suppressed, and hence the members of this second
and inner circle come between where the others
would have been.

Having stripped off the calyx, the corolla, and the stamens we come to one solitary organ in the middle —the seed-bearing organ or *pistil*, with the *stigma* at the summit.

Here I come to, perhaps, the most interesting portion of my paper, a peculiar feature in the physiology of the Primrose. Those of us who have been in the habit of not merely gathering flowers but also of examining them, must have noticed that in some flowers of the Primrose there is a small spherical body, a little larger, perhaps, than a pin's head, in the centre; that in others this organ is invisible, and instead we have deeply hued yellow bodies, five in number, clustering together. These two varieties of the Primrose are known respectively as *Pin centres* and *Rose centres*; the same peculiarity is observable in Cowslips and the garden Polyanthus. The rose centre is accounted the more handsome of the two and is cultivated to the exclusion of the pin centre. Now what are these organs? The little round body in the pin centred specimens is the *stigma*, i.e. the summit of the *pistil*, or seed-producing organ; the "rose centres" present us with the *anthers*, or pollen-bearing organs at the summit of the stamens. But if you take a rose centred specimen and dissect it, you will find the pistil inside, but reduced to about half the length, thus allowing the stamens to tower above it. So if you dissect a pin centre you will find stamens below. Such are the *facts* to be gleaned by simple observation; what I have to say now will show what is to be learned by thought and careful scrutiny from these facts. You all know that no seed ever arrives at perfection without the pistil being fertilized by pollen scattered on it from the anthers. Now what follows from this? We have first a short-styled variety, the anthers reaching some height above the stigma; it is easy enough to imagine how the pollen may drop from these anthers on the stigma and so fertilize it. But what about the long-styled pin centre, where the stigma stands above the pollen bearing organs, and so does not stand any chance of being thus fertilized? Now the first thought that struck Mr. Darwin was this, that the primrose was gradually becoming *diœcious*, *i.e.* that by and by the stamens would die out in the long-styled specimens, and the plant would bear only a pistil; and that in the short-styled specimens the pistil was dying out and the stamens were developing themselves. It is the case in many plants that some flowers bear pistils only and others stamens only—they are called *diœcious* or " two-housed." Thus in time we should get two sorts of primroses much more distinct than they are at present. Again, the thought may strike you at once, that if the pollen cannot reach the stigma on the long-styled specimens there can be no fertilization, and therefore no seed, and consequently the pin centres must in time die out. Now we must ask some botanical friend whether he knows for a certainty that the pin centres ever produce seed. He says " Yes, quite as often as the others." Therefore they do get fertilized. This sets us thinking again—how? Mr. Darwin says it is done by the Humble Bee. Many of you will recollect how by the agency of this and other insects orchids are fertilized. The bee inserting its head into the corolla of a flower detaches some of the pollen, and upon entering another is pretty certain to leave some of it behind. It is now a well-known fact that this process, which is called cross-fertilization, is the rule and not the exception; the exceptions lie on the other side, the case of a flower fertilizing itself by its own pollen, being rare even in flowers like the short-styled primroses. And when such self-fertilization takes place the seed deteriorates.

But then you will say in a case like this—the short-styled Primrose—how is it possible to prevent self-fertilization? Now we make use of another or two of Mr. Darwin's discoveries in the physiology of plants. He discovered that when a **stigma** is covered with two or three kinds of pollen,—species or varieties, one only takes effect to the exclusion of all others; also that the pollen from a long-styled Primrose is more powerful on the stigma of a short-styled specimen than its own, and *vice versa*: hence if the stigma in a rose centre gets sprinkled with pollen, both from its own anthers and from those of a pin centre, the latter will be most probably the effectual agent; and if the stigma in a pin centre receives pollen by any means from its own anthers (which is unlikely) and from a rose centre, the latter only will take effect. Hence a cross fertilization goes on, and this you see by the agency of insects. It may not have struck you, could not in fact, unless some of these facts were known to you, that insects and flowers are mutually necessary to each other, and neither could exist without the other. I remember being much struck with this remark when I heard it from the lips of Mr. Bates, the traveller of the Amazons, during a short walk on the Warren. " If insects perish " he said "flowers must necessarily perish too." The flower yields its nectar to the insect, and the insect in return assists in perpetuating the flower.

Such are a few of the mysteries enveloped in a Primrose blossom. As my remarks have already reached a greater length than I intended, I must say nothing about the Cowslip, though I should like, if I am not tiring you, to say a few words about the *Oxlip*. I may astonish you, but I do not think I shall make a rash assertion, if I say that, in all probability, none here present have seen the true Oxlip. It grows in Cambridgeshire and perhaps one or two other localities. What we call the Oxlip is, as you know, a set of several flowers like Primroses growing on one stem like Cowslip flowers. I remarked a few minutes ago that the stem of the Primrose was arrested in its growth, that if you cut through the root just below the ground you would see that the pedicels all sprang from one circle, and that if we could only imagine the stem elevated, carrying this circle with it, we should have our Oxlip. It has been generally set down as a hybrid between a Cowslip and a Primrose, but I am quite of the opinion of my friend, Mr. Britten (to whom I am in fact indebted for some of the thoughts I have placed before you), that it is but a developed Primrose. I have found both single flowered and many flowered stems growing on the same root. It is a question, however, by no means settled, whether there is not in addition a true intermediate form between a Cowslip and a Primrose. There is plenty of work before us all in the matter if we like to commence the work of observation. I chose the subject because I thought it might give us all an object to work for at once, as the Primroses are now coming out. I give you a hint or two about it. Set notes down in your vademecum (I suppose no member of a Natural History Society goes out without a note-book) to work out answers to the following questions :—

1. Are these oxlips most plentiful among primroses or among cowslips ?
2. Do those which occur among primroses bear a closer resemblance to primroses than to cowslips ?
3. Do those among cowslips resemble those flowers most ?
4. Have you ever found primroses and oxlips on the same root ?
5. Have you ever found cowslips and oxlips on the same root ?

In all probability the Primrose, Cowslip, and Oxlip have been developed from one common form, according to surrounding circumstances.

VII.—THE PUSS MOTH—A LIFE HISTORY.

Dicranura vinula.

——:o:——

Read before the Folkestone Natural History Society,
December, 15th, 1874.

——:o:——

There are Entomologists and Entomologists—two
distinct classes, better known perhaps as students
and collectors, and wide is the difference between
them. There are those who take a delight (as every
true naturalist must do) in making the lives of GOD'S
creatures a perpetual study as well as a source of
endless recreation, and I may even say amusement;
who prefer to watch animals in their native haunts
and there to trace out their habits, and their wonderful adaptation to their physical surroundings; **or who**
take advantage of the dominion given to man over
the earth and all that is in it, to capture and study
in captivity those points which probably never could
be studied in the creature's wild state. Many of
this class make no collections at all, but they are in
themselves perfect encyclopædias of all that relates
to insect life. There are others again whose sole
aim is to hunt, capture, and kill; who glory in
nothing so much as the possession of a species which
their near neighbour and friend does not possess
—whose greatest boast is that they have a unique
specimen. The habits of their so-called pets are
utterly unknown to them, and they will spend a
small fortune in purchasing specimens which they

might with very little trouble obtain by their own efforts.

Of these two classes, members of the first delight most in conversing about the objects they study, while those of the second prefer to show you their collection : the former will travel miles to see their favourites in their native spot, but you must be very cautious in imparting the locality of any rare species to the latter, for so long as they can catch or obtain specimens they think little of extermination.

There are, however, entomologists who hold a position between these two—a class I am happy to say which is increasing every day ; they make a collection, it is true, but it is quite subordinate to the study of the insects themselves. When showing this collection to you they will point with pride to such and such a species—not because they gave five or ten pounds for it, or because Mr. So-and-so has not yet been able to get one—but they tell you they caught it themselves deep in the mountain world of Wales or Scotland, and then come welcome little episodes of their travels, how up in the North they found themselves quite out of the civilized world and couldn't even get bread except the postman brought it, and that meat was utterly unknown, and that if owing to the forewarnings of a friend they had not provided themselves with a few tins of Australian meat, they might just as well have been in the backwoods of America.* Or again they point out a species they have reared from the egg, and consequently they are in possession of knowledge unknown to the rest of the world, for it may safely be said

* This was actually the case with one of my entomological friends.

that he who watches the progress of any insect, new
or old, known or unknown, from the egg to maturity
is certain to discover some fact about it previously
unknown.

There is a strange charm about this study. " Do
not you," says Kingsley in his delightful *Glaucus*,
" do not you, the London merchant, recollect how
but last summer your douce and portly head clerk
was seized by two keepers in the act of wandering
in Epping Forest at dead of night, with a dark
lantern, a jar of strange sweet compound, and innumerable pocketsful of pill-boxes ; and found it
very difficult to make either his captors or you believe that he was neither going to burn wheat ricks
nor poison pheasants, but was simply sugaring the
trees for moths as a blameless entomologist ? And
when in self-justification he took you to his house in
Islington, and showed you the glazed and corked
drawers full of delicate insects, which had evidently
cost him in the collecting the spare hours of many
busy years, and many a pound too out of his small
salary, were you not a little puzzled to make out
what spell there could be in those ' useless moths '
to draw out of his warm bed, twenty miles down the
Eastern Counties Railway and into the damp forest
like a deer-stealer, a sober white-headed Tim Linkin-
water like him, your very best man of business,
given to the reading of Scotch political economy,
and gifted with peculiarly clear notions on the cur-
rency question ? " There is, I say again, a strange
charm about this study. Perhaps one of its sources
is this constant discovery, constant acquisition of
knowledge. It is like travelling in a foreign land

while close at home—akin to the continual delight of childhood at the gradual opening of the wealth of novelty in the world around.

For myself, I shall never forget the delight with which, years ago, I watched a caterpillar undergo its wondrous transformations, when even the larva of a cabbage butterfly proved a thing of beauty; or the astonishment which filled me, not unaccompanied by reverence, as I saw the resurrection of a Peacock Butterfly from its six weeks' entombment. It had entered its shroud a creeping thing of earth, it emerged a joyous bright tenant of air—how changed and beautified. We can understand the entomologist who in his feeble old age would be wheeled to a sunny hillside and spend hours in watching the gambols of the creatures which had been his daily companions for years.

Very often our interest in entomology is first excited by seeing some creature more strange in its habits or bizarre in aspect than others we have known; it was so in my case on having (while living in the outskirts of London) several caterpillars of the Puss Moth brought to me from some willows close by—they came in company with a number of beautiful musk beetles and a thick fleshy larva of the Goat Moth with its peculiar odour. Of the latter I could make nothing; I did not know its food, and only kept it long enough to learn that it could gnaw a hole through a box like a rat. But the Puss Caterpillars ate, and grew, and their colours grew brighter and brighter, though I failed in rearing them for that time. Since then never a year has passed without my rearing boxes of caterpillars and my early friend among them.

I am afraid you are beginning to ask when I am going to talk about my subject, so without any more digressions (though I am rather fond of them) I will commence. Peculiar looking as the creature is, so that the attention of a stranger is at once drawn to it, it is really one of the very " Common objects of the country ;" here in Folkestone you may gather them by the dozen all the summer through, where-ever you come across willow or poplar tree. The Lower Road is perhaps one of the best hunting grounds.

About June or earlier we may find the brown eggs of the moth laid on the flat surface of the leaves, one here, and one there, hemispherical in shape and attached by their flat sides. With a tolerably good magnifying glass hexagonal markings may be seen on them, for the eggs of moths and butterflies, small as they are, are often beautifully sculptured—they are in their way more beautiful than those of birds. If you see a tiny hole in this brown egg it is a sign that the inmate has escaped, and probably he is close beside it, looking at you all the time, but unrecog-nised—mistaken for a bit of black cotton. He is not unlike it and looks remarkably stiff and even uncom-fortable ; almost wholly black, with a head, or at least what you suppose a head large out of all pro-portion to the rest of his body, which tapers towards the tail and ends in two fine points. Collect as many of these bits of animated black cotton as you can find and take them home, put them in a glass cylinder or tumbler, and supply them with fresh leaves from the tree on which you found them (or one of the same species), and you will have enough

H

to do to notice them for many weeks. Sometimes nibbling bits from the leaves, at others sitting motionless for hours with head drawn up and back, and tail curved over the body, and then again perhaps going a day or two without food. This is because one of the important epochs of his life has arrived—an important epoch to all juveniles—he is to have a new suit of clothes. Caterpillars are like lobsters and crabs—in fact they form a subdivision of the same class—the growth of their skin does not keep pace with that of their body ; consequently they now and then begin to feel uneasy, they haven't room, and, wiser than juveniles of a higher order, they leave off eating. By and by the old skin cracks down the back, and comes off, while underneath it appears a new one in all the glory of fresh unfaded colour. Probably as a matter of economy, the creature devours its old clo'. This you may perhaps recollect, is done by frogs, toads, and newts, though not always. A change of skin with our Puss Caterpillar occurs, I believe, four times, though other larvæ change more frequently, and at every change the markings become more definite, though in this case the colours do not always become brighter. Your attention will probably be drawn, particularly in the early youth of the larva, to two projections on the head like small ears, giving to the creature a marvellous resemblance to a young kitten ; these gradually disappear, being absorbed into the body, but their former position is marked by two black spots.

When full grown the caterpillar has a flat head, pale brown in colour, black at the sides ; when at rest it is withdrawn into the second segment of the

body, and is then surrounded by a pink line with
black spots. The general ground colour is a rather
pretty green, on the back is a very long diamond-
shaped mark of a reddish purple colour, but varying
much in different specimens, and at different ages ;
this is surrounded by a white stripe. This arrangement of colours is probably to a certain extent pro-
tective, resembling the leaf and the bark. Another
curious feature in the larva is the forked tail. It may
not be known to some of our non-entomological
friends that caterpillars in general have as organs of
locomotion six legs and ten claspers or pro-legs ; of
these the front six are retained by the perfect insect,
but the claspers never reappear after the chrysalis
state is assumed. Two of these claspers are quite at
the end of the body and are called anal claspers. But
here in the Puss Caterpillar they are not to be found
—what has become of them ? They have been
modified into the two horns we have noticed—altered
both in structure and function—one instance in
nature among ten thousand of the modification of an
organ to perform a different office—the best illustration being, I think, that of the two front limbs of
different animals, which, all formed on the same
general plan, serve as organs for climbing in the
monkeys, for flying in the bats and birds, for holding
and tearing in the carnivora, for simple support and
locomotion in the horse and elephant, and in man,
for a thousand delicate offices necessary for our existence or amusement.

Our Puss Moth derives its scientific name from
this peculiarity. Towards the end of last century,
a Bavarian zoologist denominated it *Cerura* or horn-

tail; in 1829 Latreille called it *Dicranura* or fork-
tail, a name which it still bears, though in fact both
are in use. You will ask for what purpose have the
anal claspers undergone this curious transformation,
or if a Darwinian, you will want to knew what com-
bination of physical conditions favoured its occurrence. Well, these questions have not been answered
as yet, at any rate satisfactorily; here is another
opportunity for any of you to make the discovery
by careful watching and experimenting. Out of
these horny cases, as they really are, the creature
can project tiny scarlet threads to some considerable
distance; this it will always do if you disturb or
irritate it by touching it, or if another larva does so.
It has been supposed that these threads serve as a
kind of protection—weapons of defence—against
small enemies, particularly the ichneumon flies to
which we shall presently refer. And when the
peculiar attitude of the larva is assumed, the head
thrown up and back, and the tail recurved so as to
throw the whole body into a concave form, these
threads can sweep over it all, so that there is considerable probability in the conjecture. But, on the
other hand, these larvæ do not seem to be more
exempt than others from the successful attacks of
these and other parasites. So the question remains
undecided.

The caterpillar has, however, another mode of
defence by way of attack, namely, the ejection of a
fluid acrid in its nature, and causing a slight smart-
ing on the skin, but probably more powerful when di-
rected against minute enemies. This power is possessed by some beetles among other creatures; I remem-

ber when coming home once in the evening, seeing a
beetle running across the road, and pouncing upon it
as my lawful prey, I picked it up, and while holding
it in the twilight between my eyes and the sky to
look at it I suddenly received a discharge below the
eye that caused for the minute some little amount of
pain, and I parted with the beetle much sooner than
I had intended. Our caterpillar sends the fluid from
a transverse slit just beneath the chin, or beneath
where it would be if he had one. It is said in some
books to lose this power in captivity, but this is not
always the case. In connexion with the tail I ought
to have mentioned that you generally find the full-
grown caterpillars possessing merely the stumps,
and no scarlet thread is ever produced. It might be
thought perhaps that this was the effect of old age,
and was similar to creatures of another class growing bald or losing their teeth with advancing years.
Let me assure you that such is not the case. For
though the larva itself may appear old we must
remember that it has yet to go through a second life,
and can, therefore, be scarcely considered in the sere
and yellow leaf. The mystery is solved by watching
the animal in captivity, when it will be seen that if
any one caterpillar can possibly catch another asleep
or otherwise off his guard he will assuredly nibble
off his tail. They eat each other's tails, and that
purely out of choice, for they do it in the presence
of an abundance of fresh food. I believe, however,
they do their utmost to prevent others returning the
compliment. Monkeys also are fond of tail ends,
but then they nibble their own,

When about six or seven weeks old the larva of

the Puss Moth is full fed and becomes much lighter
in colour; it then begins to seek a suitable place on
which to construct its cocoon in which to spend the
winter. It generally fixes on the bark of a tree, and
fashions its cocoon out of the said bark. With a
gummy secretion of its own it fastens together small
chips which it has gnawed from the bark, thereby
making a hollow, building all round itself and arch-
ing a covering over it just as the Esquimaux builds
his snow house, only that it does not leave any
aperture for escape. It requires the quick discern-
ing eye of the practised entomologist to detect one of
these cocoons on the stem of a tree. It is similar
in colour to the bark, and is merely one of a great
number of excrescences upon it. I have often
hunted for them in vain, but they were plain enough
after the escape of the moth, the large aperture and
the white down round it making it conspicuous.
This is another instance of what is called "pro-
tective resemblance," and saves the creature from
the attacks of birds that live on larvæ and pupæ and
whose quick eyes soon discern anything differing in
colour from its surroundings. In the "struggle for
existence" always going on, the Puss Moth thus
stands a better chance of preservation, a less risk of
being exterminated. The cocoon is very hard, and
a penknife has often been broken in the attempt to
cut or remove it. In captivity of course the creature
has not much choice, and will make its cocoon of
anything it can get; if confined under a glass it
will even construct it simply of gum. The case is
then transparent and the brown chrysalis can be
seen through it. In the specimens before you there

are two or three cocoons formed on trees, and one formed in the mortar of a wall. This last is very peculiar, and it is difficult to account for the creature's choice when trees were within six feet of the wall. In fact the caterpillar appears to have descended a poplar tree close by and to have gone across to the wall. The other cocoons are of course natural, but were made under circumstances over which the larva had no control—one is formed on and out of a piece of coal sacking, the other of coloured bits of cardboard. But one and all are weatherproof and watertight ; no storm, however violent, disturbs the chrysalis within.

Out of this, in May or June, comes the perfect moth or imago ; it moistens its hard covering with a solution manufactured on the premises, and so makes a hole large enough to crawl out by. Not yet, however, does it appear in full perfection. The body and wings hang down limp and almost helpless, the latter being not more than a quarter of an inch in length. But it crawls upward and suspends itself for an hour or two by its legs, and then you may almost see the wings grow and strengthen. One specimen on the table shows you its state on first emerging. It never succeeded in enlarging its wings, owing to some inherent defect. The perfect insect is not at all lacking in beauty, thickly clothed with down to the very toes, and with lighter and darker markings on the wings, which are supposed to resemble those of a variegated cat, and hence the name of Puss Moth. We have seen, however, that the kitten-like aspect of the young larvæ, may also have something to do with the name.

You may perhaps be disappointed now and then

in not getting any moth to emerge from some of your cocoons. On breaking them open you will find a cluster of long black cases inside. These are cocoons out of which have escaped ichneumon flies —the flies and their cocoons are both on the table. You ask how came they there? And where is the chrysalis? All caterpillars are liable to the attacks of these creatures, the office of which appears to be to assist in preserving the balance of nature by checking the undue increase of vegetable-feeding caterpillars; in this of course they act in concert with the birds. You will generally find that in those years when caterpillars are more than ordinarily abundant, when the gardener finds his cabbage plants worked into verdant antimacassars of every complicated pattern, that then ichneumon flies are also abundant, and that comparatively few of the caterpillars arrive at butterfly-hood. I once counted nearly 100 chrysalises of the Cabbage White Butterfly in an outhouse, and with the exception of not more than half a dozen, each one had beside it a cluster of small yellow cocoons—those of an ichneumon fly. This fly darts down on the unfortunate caterpillar's back, bores a hole in it and lays therein an egg; this egg hatches and the small larva feeds upon the bodily substance of its huge brother, carefully avoiding, it is said, the vital parts, " growing with his growth and strengthening with his strength."

It would not appear as if much inconvenience resulted, for the unwilling landlord eats and goes about his allotted work with all unconcern, and often has strength sufficient at last to manufacture the tomb out of which he is destined never to arise.

VIII.—ON THE CLOUDED YELLOW BUTTERFLY AND ITS ABUNDANCE IN 1877.

———:o:———

Read before the Folkestone Natural History Society, May 14th, 1878.

———:o:———

Those of our members who take an interest in entomology have not even yet, I venture to say, forgotten the sight which gladdened their eyes during last summer from June to October. Even our non-entomological friends, who, without devoting themselves to the exciting influences of the chase can yet notice intelligently all they see round them, well remember the hundreds of Clouded Yellow Butterflies which covered the slopes of our Folkestone hills. For it was the great " *Edusa*" year, and as such will be long remembered. Never before within the recollection of any one was there known to be such an abundance of this handsome insect. All day long from the time the sun began to chase away the dew to the time it fell again, field and meadow and hill side were animated by hunters with the net ; they had never had such a chance before, they never expected to get it again, and in very deed they made the most of it, for captures by single individuals were recorded by hundreds.

The Clouded Yellow, or *Colias Edusa*, as it is scientifically termed, is one of those butterflies that are very capricious in their times of appearance. For

one or two years, or even longer, we may not see a
single specimen ; another year they come out by twos
and threes, but every now and then there is a great
swarm of them.　No one can yet account for it ; no
one knows whether it is through extensive immigration, or through circumstances more than usually
favourable to their multiplication on our own soil.
It is one of the many knotty but interesting questions
Nature sets before us, and almost defies us to answer.
Perseverance, however, and patience on our side will
doubtless in the end write down a correct answer to
this, as has been done to other problems.　A few
years ago (in 1872) we had a "Camberwell Beauty
Year," when this large insect, noted as one of the
more than usually rare British butterflies, and indeed
generally referred to as rapidly disappearing like the
Great Copper, was taken in many parts of the country, though not, it is true, in any great numbers.
Folkestone, however, yielded several specimens.　The
Bath White and the Queen of Spain were abundant
the same year.　And so it is with other insects from
time to time.

As *Edusa* is of great local interest, I thought a few
remarks concerning the butterfly, its appearances,
and its variations in form and colour, would not be
unacceptable to our members.　By the kindness of
Mr. Giles and Mr. Blackall we have before us some
fine specimens as illustrations.　The family to which
it belongs is known as the *Rhodoceridæ*, or Redhorns,
from the roseate tinge you will observe on their
antennæ.　It is not a large family in our island, including only the following members : The Clouded
Yellow (*Colias Edusa*), the Pale Clouded Yellow (*C.*

Hyale), and the Brimstone, (*Gonepteryx Rhamni*).   The caterpillars belonging to them are cylindrical in shape, more like what is called a grub—smooth and without any spines.   Those of the first two feed on leguminous plants, those of the latter, as the specific name implies feed on the Buckthorn (*Rhamnus frangula* and *R. Catharticus*).   In the perfect insect the ground colour is yellow, varying from the rich orange of *Edusa* to the delicate greenish yellow of the female of *Rhamni*. The markings are orange and black. All three insects are strong and rapid fliers. They probably all hybernate through the winter either regularly or occasionally ; certainly the Brimstone does, for the first warm days of the year generally supply a few stray specimens. One is astonished at the marvel- lously fresh and unfaded appearance of some of these individuals ; it is difficult to imagine where they can have sheltered themselves so securely for their winter sleep.   I have caught them in March without a scratch or a mark of any kind on the wings, and with the beautiful gray silky comb of hair on the thorax perfect and unruffled.   We also have evidence of hybernation on the part of *Edusa*.   Mr. Arthur Taylor showed me, on the 28th of April, 1869, a specimen he had just taken on the Warren : it was small, tolerably fresh, and had undoubtedly survived the brood of the year before. We have already had notices of the reappearance of last year's specimens, though I am not aware that any have been seen at Folkestone. The proper time of its appearance fresh from the chrysalis is in July, perhaps even the latter end of June, and it was formerly supposed that this was the only brood in the year. Then the appear

ance of some late but very fresh specimens in
October led entomologists to conclude that there were
two broods. It was not, however, till last year that
any evidence was obtained proving that their might
be in some years three. Nowhere was this proved
more satisfactorily than in Folkestone, by one of our
own members, Mr. Blackall, who has kindly placed
at my disposal notes of his observations on the insect
throughout last summer. The first appearance was
in June, abnormally early. On the 9th of that
month as I was in a field by Lady Wood, a specimen
of *Edusa* rushed by me in its well-known headlong
manner. Those who have ever hunted it know what
a chase it is. The day was hot, I did not want to
run about, and like the fox in the fable, though I
should have liked to capture it, I said to myself "It
is but a hybernated specimen." Please give me the
credit, however, which we do not award to the afore-
said fox, of believing what I said to be true. Presently
a second flew by, then a third, and a fourth. I
caught the seventh, which was in such good condition
that I determined to have as many as I could get.
Of course after that determination I saw no more.
On mentioning this to Mr. Blackall, I found that he
having more opportunities than myself, had already
taken some specimens, the first being captured June
6th. We both visited the field I had been in the
next morning, but only saw one. Within the next
few days dozens might have been taken; many in
perfect condition; one, in fact had its wings still
limp. There was therefore no doubt about their
being freshly emerged specimens; probably none
had hybernated. Some of these were of unusually

large size. Mr. Blackall took one of the white variety known as *Helice*, to be alluded to again presently; this was much worn. We were puzzled rather about this appearance of *Edusa* at the beginning of June, apparently between the period when hybernated specimens might be caught, and that when new broods occurred, and together with Mr. Giles, who had also been working, we pondered over the question how they were to be regarded. At one time we said " hybernated," and again we said " no." Anyhow, we all prophesied a second brood, probably numerous, at the latter end of July or beginning of August. Accordingly, on the third of the latter month they swarmed, and *Helice* among them, to the intense delight and astonishment of everyone. From this time onward for nearly four months the whole neighbourhood abounded with them, the fading specimens of the second brood mingling with the fresh bright ones of a *third* in October. It is this third brood which, so far as I am able to make out, is a unique occurrence; a second brood has been recorded, though doubts were thrown upon it, but no one expected a third. The weather being favourable, and Mr. Blackall very energetic in the matter, specimens were captured all through October and up to 12th of November. Some of these in October he took while they were drying their wings after emerging from the chrysalis. Many of this third brood proved to be much below the average size. In one of the cases on the table the three distinct broods are shown for comparison. *Helice* was abundant again in the last brood. Mr. Blackall took ten between the end of October and November 12th.

Such is a brief outline of the occurrence of this insect in our own neighbourhood. But the same tale might be told of almost every part of England. One correspondent in the *Entomologist* knew of 1,500 captures in the district round Strood. The third brood made provision for the present year, though some we know hybernated. My friend took eggs off the trefoil in November, hoping to rear some specimens ; in this, however, he was unsuccessful. I might mention here for the benefit of entomologists a " wrinkle worth knowing" with regard to the mode of capturing *Edusa*. I have it again from Mr. Blackall's notes. Don't wait till mid-day when the hot sun inspires the insect with unusual vigour, and correspondingly depresses your own physical energies. I assure you it is no joke racing over the Folkestone Downs after it then, and especially when you have to go *up* hill. You think perhaps that is a feat you would not attempt for the sake of a specimen. Ah, wait till you see *Helice* go by ; and with a loud exclamation off you go ; up or down it is all one, you feel you must have *that* specimen. It is to be hoped you may get it, but whether you do or not you will want ten minutes rest afterwards. The wise plan is to go out early in the morning when the sun has just, *only just*, driven off the dew. *Edusa* is rousing itself, and rising lazily only to drop again a short distance ahead. Mark the spot, walk up quietly and put your net over it.

And now we come to ask how we are to account for the great abundance of this butterfly in 1877—a year which has been alluded to as " one of the wettest and most sunless years remembered for some time ?"

Not by immigration certainly; I have given satisfactory proof that the specimens were British born.
That they do, like some other butterflies, occasionally
come across from France we know full well; clouds
of them have been seen on their way far out at sea.
There is no explanation forthcoming yet, we must
extend our knowledge of *Edusa* as regards its life
history both here and on the Continent; we must
accumulate facts and patiently await the answer to
our question. There was an old tradition that it
occurred in England once every seven years; this
empirical law was founded on scanty observation, and
has quite broken down before stubborn facts. We
are all looking forward with anxious expectation to
the *Edusa* period of 1878. Will it be one of plenty
or scarcity? If the latter, why? It is very remark-
able that the Pale Clouded Yellow (*C. Hyale*) was
very rare indeed last year, though in 1876 it was
tolerably plentiful. Why should it not have occurred
as well as *Edusa?* Closely allied as the two are,
both in the same genus, both living on the same or
similar food, what are the circumstances in nature
that caused one to be abundant, the other absent?
Eggs in abundance must have been left the previous
year; we have always supposed the same weather
good for both; where then was *Hyale* in '77? We
cannot tell. You see I am asking plenty of questions,
and giving no answers to them. My aim is to show
what interesting problems there are to be worked out
in connexion with these butterflies, and that it is
in the power of anyone of us to assist in solving them,
by a little patient study of nature. I promised to
allude again to the white variety of *Edusa* known as

Helice. There are several specimens on the table, varying in size and markings. Now a few years ago *Helice* was particularly rare; I well remember the capture of my first, ten years ago. I was driving out with a friend through Acrise when I believed I saw *Helice* settle on a flower by the road side. My friend kindly stopped while I went back, *Helice* as kindly waited to be caught, and I came back with it in my box. It was a red letter day with me. But last year it was not much thought of; it came into our gardens; Mr. Austen took 78. I suppose there is scarcely a cabinet now without it. Now if *Helice* occurred in a separate country and away from *Edusa*, there is little doubt that it would be set down as a different species altogether, and not as a mere variety like a white hyacinth. There are many species that differ far less than these two. Stay; there is one fact against making *Helice* a species—every specimen is a female; there are no white varieties of the males. You will agree with me that this is a very curious fact. Why should the female *Edusa* occasionally be white instead of yellow?

"And yet more why"—

Why should the male never undergo such a change? Again I have no reply, only a suggestion. But first let me state that the occurrence of two forms of females with only one of males is not restricted to *Edusa*. It is known to exist with several other butterflies. Mr. Wallace, in his book on the Malay Archipelago, alludes to a very striking instance in *Papilio Memnon*, a large handsome butterfly of a deep black colour with various coloured markings, measuring five inches in expanse of wing. There are two forms of females, the difference between them being much greater than that between *Edusa* and *Helice*. One possesses two tail-like appendages on the hind wings, of which neither the male nor the other female possesses the slightest vestige. The colours too are different. The tailed variety however closely resembles the normal female of an allied species, *P. Coon*, so that we have here a case of mimicry. In the north of India *P. Coon* is replaced by another species *P. Doubledayi*, and *P. Memnon* is replaced by an allied species *P. Androgeus*. And here curiously enough there are two female forms of the latter, one of which mimics the female of *P. Doubledayi*. Mr. Wallace offers the following in explanation :—" The butterflies imitated belong to a section which from some cause or other are not attacked by birds ; and by so closely resembling these in form and colour the female of *Memnon* and its ally also escape persecution." If so, then I should imagine that the tailed variety will prove permanent, while the other will die out. All three forms of the male and the two females are produced from eggs of either female. Mr. Wallace's explanation serves to show theoretically how one form may replace another. Well now, in the case of *Edusa*, have we here an example before our eyes of a transmutation of species or of form gradually going on ? Will this change some day extend to the males, and will there be by and by a totally distinct species having a white ground colour ? or will the change be confined to the females, the old form of which will die out ? In connection with this again, does the white colour of *Helice* afford it any protection against enemies ?

I

Is it mimicry of *Hyale* or of any of the Whites ?
Do birds devour the yellow specimens while white
ones for some cause or other are passed by ?   Or on
the other hand do birds devour white specimens and
leave the yellow ones which are so much in excess
of the others ?   Is the white variety only a reversion
to the original colour ?   More questions again for us
to puzzle our minds over.

You may perhaps say that from a bright orange
ground colour to a white can hardly be looked on as
a *gradual* change as we see it.   Last year's captures
have enabled me to reply to this.   Mr. Austen
showed me a specimen he had taken in which the
ground colour was a beautiful light golden yellow ;
and it was very doubtful as to whether it ought to be
labelled *Edusa* or *Helice*.   Mr. Blackall has a similar
specimen, and I have one lighter still, but not white.
I then ventured an opinion that if it were possible
to place side by side a very large number of captures of
the two forms we should find every tint, from the orange
of *Edusa* suffused with a rosy purple light to the milk
white *Helice*.   Such I find was the case at the Entomological Exhibition, held in March at the Royal
Aquarium.   It was absolutely impossible to say where
Edusa terminated and *Helice* began.   In the plate of
examples on the table you will also notice this.   And
in addition you will find one specimen having the
fore wings of *Helice* with the hind wings of *Edusa* ;
another again which has both wings on the right side
those of *Helice*, while the left are those of *Edusa*.
So you see it is possible to believe in a *gradual* transmutation from one to the other ; but it is impossible
to say why the change is limited to the females.

From the eggs of *Edusa* both yellow and white specimens have been reared ; this of course would be a satisfactory proof of their not being distinct species. With regard to the variation in markings, it has been noticed that we had several illustrations last year of specimens closely resembling continental ones bearing a different specific name. When the full history of the genus comes to be written, it will doubtless be a deeply interesting one ; and will tend to show the great probability that many so called species are but variations of one original form, such variations being attributable to differences in food, climate, and other physical surroundings.

IX.—VIOLETS.

Read before the Folkestone Natural History Society,
at Saltwood, June 6th, 1879.

———:o:———

"Spring Violets!" says Miss Pratt in her
"Flowering Plants of Great Britain "; "What lover
of the country is not gladdened by their coming?"
I leave the question to be answered by those whose
hearts have been gladdened by the balmy (?) days of
the late spring, if they have noticed such a season.
True the Violet is essentially a spring flower, associated with verdant banks, young lambs, and the
warbling of birds under a blue and sunny sky; at
least so we have been taught. I suppose such things
have been in the past, in the good old times, just as
there have been fairies and Arabian Nights. Now,
alas! all are fled.

When I first thought of writing a few lines about
Violets, I felt in duty bound to keep up old traditions,
and had already made a choice selection from poets
and others of suitable phrases, such as " the velvet
foot of Spring," " balmy breeze," " vernal airs,"
" the kisses of the April wind," &c., &c. I sat
down—it was on the ninth of May in the present
year of our nineteenth century—the century of pro-

gress and improvement, mark you, in all that appertains to the well-being and comfort of man ; on the ninth of May, 1879, I took my pen to write in praise of Violets, and before I had written six words there came on a blinding snowstorm. What could I do with such a scene before me ? Need it be said that all those beautiful phrases fled, leaving a blank, and an aching void behind ? What kind of a paper I might otherwise have written, what the loss to my present audience may be through that fall of snow, it is not for me to say. Let me only beg those who may be disappointed in my paper, to attribute its shortcomings to the chilling influences which have been at work around me. I hope they will at least acknowledge that it is a difficult task to write about Violets while a snow storm is in full action.

Our faith in bright spring days is gone ; nevertheless Violets do occasionally bloom, and doubtless in the intervals between the snow storms some of us have paid a few delicate attentions to them during the late conclusion of winter.

I have been anxious to discover how many species we have in our own district, and if possible to learn the distinguishing marks between them. Let me first give a general description of the flowers, and then enumerate the various species.

They belong to the natural order which takes its name from them—*Violaceæ*. It includes twenty-one genera, one only of which, *Viola*, is found in the British Islands. The parts of the flower are arranged in fives, that is to say—there are five sepals, five petals, and five stamens, the last mark referring the plant at once to the Linnæan class *Pentandria,*

There is a curious modification in these stamens; the summits are expanded into flat membranes, which bend towards the centre, making a small cavity into which the pollen falls when ripe. Two of these stamens send out long processes behind known as "anther spurs," which are enclosed in the spur formed by the lower petal of the corolla. These anther spurs possess glands which secrete honey, and it is not at all uncommon to find the corolla spur perforated in one or two places where insects have been endeavouring to regale themselves. What these insects may be I do not know, but I believe moths and bees in their visits to the flower obtain the nectar in the usual way, namely, by passing the sucking tube down the throat of the flower from the front. The petals of the flower are unequal in size, the lower one being lengthened into the spur just mentioned. The general colour is a variable purple, sometimes lilac, or white, and in one or two species yellow. In the throat of the flower are two tufts of hair, one on each lateral petal; they possibly protect the entrance from rain which might enter and wash out the nectar, though this in the great majority of cases would be prevented by the drooping posture of the blossom; probably also they guide bees or other insects in their search after honey; they certainly add greatly to the beauty of the corolla.

It is not generally known (except to the botanist) that the Violet, like some other plants, has two distinct and quite different sets of flowers; the second set, small and inconspicuous, appearing about six weeks or later after the first. They are known as "cleistogamic" flowers; they are uncoloured, and

almost without petals, but they produce the seed
which preserves the species ; very rarely is this perfected by the ordinary blossoms, in fact it has been
said by Bentham that the pansy (*V. tricolor*) is the
only one of our English species in which the showy
flowers generally produce seed. The cleistogamic
flowers are self-fertilizing, and therefore do not require the visits of insects to aid in the perfecting of
their seed. They are believed to be degraded forms
of the ordinary showy flowers, since all intermediate
forms can be traced. If this be the case, it seems
to me that the continued existence of the ordinary
coloured flowers is difficult to be explained on Darwinian principles. According to his theory (which is,
as I have shown in a former paper, that generally
accepted by naturalists), the colour of flowers has
been developed by the visits of insects, and such
colour has been developed only because it was ad-
vantageous to the plant in the " struggle for existence." But if, as is believed, the cleistogamic flowers
are aborted forms of the showy ones, then it must
have been a great advantage to the plant to possess
such aborted forms. And if the violets are as a rule
perpetuated by the development of seed from these
colourless blossoms, unvisited by insects, of what
advantages to the plant at all are the coloured
flowers ? It could get on just as well without them.
Sir J. Lubbock says he thinks " the persistence of
the showy ones can only be accounted for by the
fact that the ordinary flowers are useful in securing
an occasional cross." This seems to me unsatisfac-
tory, and implies an enormous excess of energy over
the effect to be produced.

The task of distinguishing between the different species of violets is a difficult one, and in fact it is by no means easy to say how many there are. To ordinary mortals, guiltless of botanical lore there are but two, viz., the Sweet Violet, and the Dog Violet. Why " dog " violet it is hard to say ; it seems a term of contempt, often applied to a thing which will not bear comparison with its rivals. We have it in dog rose and other flowers. The epithet " horse " is similarly prefixed in horse-radish, horse-mustard, &c. Both terms are found again in connexion with the names of animals, e.g., dog-whelk, dog-fish, horse-mussel, horse-stinger (the dragon fly), &c.

But the species of violet are by no means limited to the above two; the name " Dog violet " usually includes all those that have no scent, of which there are certainly several well marked forms All botanical authorities vary in their number of species. Dr. Hooker in his Student's Flora gives five, including several sub-species under *canina;* Miss Pratt gives seven, identical with those of Withering; Babington gives eight. I suppose there is no house, society, or sect, where absolute unity of views on any particular subject is to be found. Botanists are certainly no exception, they are divided into sections known respectively as *Lumpers* and *Splitters;* the former reducing the number of species as much as possible, the latter increasing them at their own sweet will. This is due to the fact that nobody is able to define what a species is, at any rate to the satisfaction of anyone else. Until that is done the two factions in botany are bound to exist. If I were giving notes for the assistance of young botanists, I should feel inclined to follow the classification given by Miss Anne Pratt as the simplest, viz., into two sections:—

(a) Those having no evident stem.

(b) Those with very evident stems.

Under the former—those in which the stem is apparently absent—are included three species easy to distinguish :—

1. The Sweet Violet (*Viola odorata*) flourishing in a few localities near us. Mr. Hy. Birch found a bank covered with them near Castle Hill.

2. The Hairy Violet (*V. hirta*) without scent, and commonly referred to the Dog violet group. It is recognised by the rough hairs covering the leaves and leaf stalks, and two small leaf-like organs called bracts are found on the flower-stalk always *below* the middle; the same organs exist in the Sweet Violet, but are *above* the middle of the stalk. *V. hirta* is the prevailing form at the East Cliff end of the Warren, and is generally distributed elsewhere ; it appears fond of a chalky soil.

3. The Marsh Violet (*V. palustris*), paler in colour than violets usually are,.sometimes lilac, very abundant in marshy spots in Scotland, and the mountainous districts of England. Occasionally it is scented.

In all these three species the green leaves enlarge considerably for some time after flowering, a pecu- liarity which most of us have doubtless noticed in the garden violets. In the second section—violets with an evident stem which continues to lengthen as it throws off flower stalks, are included four species :—

1. The Dog Violet (V. *canina*). Here our

troubles begin. What is a Dog Violet? Under this species Dr. Hooker includes the Wood Violet (*V. sylvatica*) as a sub-species; the form is common enough here. But Babington makes *sylvatica* a distinct species, and Miss Pratt does not mention it by name, though she evidently refers to it in some of her remarks about *canina*. The difference given is that in *canina* there are primary and lateral stems, *all* giving off flowers and lengthening; while in *sylvatica* the lateral stems only do this. It appears to me that the distinction is an uncertain one, and very difficult in many cases to detect. I wrote to a friend in the British Museum to ask if he could give me any definite marks for a guide, knowing that he and I had been sorely exercised in the matter in our rambles over the Buckinghamshire hills. I look upon him as an authority, but he only tells me he cannot define *canina*, though he knows it when he sees it; so I have not gained much. I found a short time ago a good specimen of—well, I cannot say the species, but it is not *odorata*, and it is not *hirta*; it has two lateral stems lengthening and flowering, the primary stem lengthens of course, but gives off no flowers. I suppose it is *sylvatica*.

This form (the Wood Violet) is by far the handsomest of all our violets, growing frequently from six to eight inches high; the flower is very large, and its colour often remarkably rich. It thrives best in woods and on slopes, and is very common round Folkestone. For the benefit of those inclined to range themselves under the banner of the splitters, I may say that Babington not only makes *sylvatica* a true species, but gives two varieties of it which he says are also probably distinct. They rejoice in the names of *Reichenbachiana* and *Riviniana*. The main differences are :—

Riviniana has many branched veins on the lower petal, and the spur is yellowish.

Reichenbachiana has a few simple veins, and the spur is purple or lilac.

I have found to my confusion that these characteristics frequently cross each other, and that branched veins, oftener than not, go along with a purple spur.

I cannot help thinking that these two forms, and also *canina* simply depend on locality, on the nature of the spot where they grow, and that they vary accordingly, being more or less robust and luxuriant as the surrounding conditions are more or less favourable. I fear many of the specific descriptions of plants are made with undue haste, and from examination of a limited number of specimens. A very large number ought to be carefully looked over, including some from every kind of locality where the flower is known to occur, before any description is ventured upon ; if this were done I believe we should certainly get fewer species. The remaining species in this section are :—

2. The Cream coloured Violet (*V. lactea*) found on mountains and boggy heaths, given by Miss Pratt as distinct, but by Babington and Hooker as a sub-species under *canina*.

3. The Pansy (*V. tricolor*), which is too easily recognised to require any special description. It is common in cultivated fields ; and from it all the varieties of our garden pansy have been produced.

4. The Yellow Mountain Violet (*V. lutea*), much resembling *V. tricolor* but usually larger. It is sometimes purple instead of yellow.

We possess then in our own district four species:—

The Sweet Violet (*V. odorata*).

The Hairy Violet (*V. hirta*).

The Wood Violet (*V. sylvatica*) of both forms.

The Pansy (*V. tricolor*).*

Whether *canina* is here or not I cannot venture to say until I have seen an authentic specimen. It is said to grow in sandy and peaty places.

Of the virtues of the Violets I do not know that there is much to say; but their roots appear generally to possess medicinal properties. One of the ipecacuanhas is the root of a Brazilian violet; in South America most of the family are shrubs.

According to old Gerard the study of violets is productive of good moral results, and in his words I will conclude. Speaking of the Sweet Violet he says "They have a great prerogative above the other violets, not only because the minde conceiveth a certain pleasure and recreation by smelling and handling these most odoriferous flowers, but also for that very many by these violets receive ornaments and comely grace; for there be made of them garlands for the head, nosegaies, and posies, which are delightful to look upon, and pleasant to smell to, speaking nothing of their appropriate virtues; yea, gardens themselves receive by these the greatest ornament of all cheerful beautie, and most gallant grace. And the recreation of the mind which is

---

* *Viola palustris* is recorded in an old Floral Guide to East Kent, as occurring at Sellinge.

ta'xen thereby cannot but be very good and honest,
for they admonish and stir up a man to do that
which is comely and honest; for flowers through
their beautie, variety of colour, and exquisite forme,
doe bring to a liberal and gentlemanly mind the
remembrance of honestie, comelinesse, and all kinds
of virtues. For it would be an unseemlie · thing
for him that doth look upon and handle faire and
beautiful things, and who frequenteth and is con-
versant in faire and beautiful things to have his
mind not faire but filthie and deformed."

LISTS

OF THE

PLANTS, BUTTERFLIES, MOTHS, BIRDS,

AND

LAND AND FRESH WATER SHELLS,

TO BE FOUND

In the Neighbourhood of Folkestone, within a radius of six miles from the Town Hall.

I.—PLANTS.

RANUNCULACEÆ

Clematis Vitalba
Anemone nemorosa
Myosurus minimus
Ranunculus fluitans
,, peltatus
,, Baudotii
,, Drouettii
,, tricophyllus
,, Ficaria
,, Flammula
,, auricomus
,, acris
,, repens
,, bulbosus
,, hirsutus
,, sceleratus
,, arvensis
,, parviflorus
Caltha palustris
Helleborus viridis
Delphinium Ajacis
Aconitum Napellus

NYMPHÆACEÆ

Nuphar lutea

PAPAVERACEÆ

Papaver somniferum
,, Rhœas
,, Argemone
,, hybridum
Glaucium luteum
Chelidonium majus

FUMARIACEÆ

Fumaria officinalis

CRUCIFERÆ

Cakile maritima
Crambe maritima
Raphanus Raphanistrum
Sinapis arvensis
,, alba
,, nigra
,, tenuifolia
,, muralis
Brassica oleracea
Sisymbrium officinale
Erysimum Alliaria
Hesperis matronalis
Cheiranthus Cheiri
Cardamine pratensis
,, hirsuta
Arabis thaliana
Barbarea vulgaris
Nasturtium officinale
,, sylvestre
Draba verna
Alyssum maritimum
Thlaspi arvense
Capsella Bursa-pastoris
Lepidium Draba
,, Smithii
,, campestre
Senebiera Coronopus

RESEDACEÆ

Reseda lutea

K

Reseda Luteola
CISTACEÆ
Helianthemum vulgare
VIOLACEÆ
Viola odorata
,, *var.* alba
,, hirta
,, sylvatica
,, a. Riviniana
,, b. Reichenbachiana
,, tricolor
,, *var.* arvensis
,, palustris
DROSERACEÆ
Drosera rotundifolia
POLYGALACEÆ
Polygala vulgaris
FRANKENIACEÆ
Frankenia lævis
CARYOPHYLLACEÆ
Dianthus Armeria
Silene inflata
,, maritima
,, nutans
,, *var.* paradoxa
Lychnis vespertina
,, diurna
,, Flos-cuculi
,, Githago
Mœnchia erecta
Cerastium glomeratum
,, triviale
,, tetrandrum
,, semidecandrum
Stellaria media
,, Holostea

Stellaria graminea
,, uliginosa
Arenaria trinervis
,, serpyllifolia
Sagina maritima
,, procumbens
,, nodosa
Spergula arvensis
Spergularia neglecta
,, *var.* media
,, marginata
,, rubra
Saponaria officinalis
ILLECEBRACEÆ
Scleranthus annuus
PORTULACEÆ
Montia fontana
TAMARISCACEÆ
Tamarix anglica
HYPERICACEÆ
Hypericum Androsæmum
,, calycinum
,, perforatum
,, tetrapterum
,, humifusum
,, hirsutum
,, elodes
MALVACEÆ
Malva moschata
,, sylvestris
,, rotundifolia
Althæa officinalis
Lavatera arborea
LINACEÆ
Linum catharticum
,, angustifolium

GERANIACEÆ

Geranium pratense
,, pyrenaicum
,, molle
,, dissectum
', columbinum
,, Robertianum
Erodium cicutarium
Oxalis Acetosella

ILICACEÆ

Ilex aquifolium

CELASTRACEÆ

Euonymus europæus

RHAMNACEÆ

Rhamnus catharticus

SAPINDACEÆ

Acer campestre
,, Pseudo-platanus

LEGUMINIFERÆ

Ulex europæus
Genista anglica
Sarothamnus scoparius
Ononis spinosa
,, arvensis
Anthyllis vulneraria
Medicago sativa
,, lupulina
,, maculata
,, denticulata
,, minima
Melilotus officinalis
,, arvensis
,, vulgaris
Trigonellaornithopodioides
Trifolium repens
,, subterraneum
,, pratense

Trifolium arvense
,, scabrum
,, striatum
,, glomeratum
., suffocatum
,, fragiferum
,, procumbens
,, minus
,, filiforme
Lotus corniculatus
,, major
Astragalus glyciphyllos
Ornithopus perpusillus
Hippocrepis comosa
Onobrychis sativa
Vicia hirsuta
,, tetrasperma
,, cracca
,, sativa
,, angustifolia
,, *var.* Bobartii
,, lathyroides
,, sepium
,, bithynica
,, *var.* angustifolia
,, sylvatica
Lathyrus Nissolia
,, pratensis
Orobus tuberosus

ROSACEÆ

Prunus Padus
,, Avium
Spiræa Ulmaria
Agrimonia Eupatoria
Poterium Sanguisorba
Alchemilla arvensis
Potentilla anserina
., reptans

Potentilla Tormentilla
,, Fragariastrum
Fragaria vesca
Rubus cæsius
,, fruticosus
Geum urbanum
Rosa rubiginosa
,, canina
Cratægus Oxyacantha
Pyrus malus

LYTHRACEÆ
Lythrum Salicaria

ONAGRACEÆ
Epilobium angustifolium
,, hirsutum
,, parviflorum
,, montanum
,, tetragonum
Circæa lutetiana

HALORAGACEÆ
Hippuris vulgaris
Myriophyllum verticilla-
tum
Callitriche verna
,, platycarpa

CUCURBITACEÆ
Bryonia dioica

GROSSULARIACEÆ
Ribes rubrum
,, Grossularia

CRASSULACEÆ
Sedum Telephium
,, anglicum
,, acre
Cotyledon umbilicus

SAXIFRAGACEÆ
Saxifraga oppositifolia
,, tridactylites
Chrysosplenium oppositi-
folium
,, alternifolium

UMBELLIFERÆ
Hydrocotyle vulgaris
Sanicula europæa
Conium maculatum
Apium graveolens
Sison Amomum
Bunium flexuosum
Pimpinella Saxifraga
,, magna
Ænanthe fistulosa
,, pimpinelloides
,, Lachenalii
,, crocata
,, Phellandrium
Fœniculum vulgare
Silaus pratensis
Crithmum maritimum
Angelica sylvestris
Pastinaca sativa
Heracleum Sphondylium
Daucus Carota
,, var. maritimus
Torilis Anthriscus
,, nodosa
Scandix Pecten-Veneris
Anthriscus sylvestris
Smyrnium Olusatrum

LORANTHACEÆ
Viscum album

ARALIACEÆ
Hedera Helix

CORNACEÆ

Cornus sanguinea

CAPRIFOLIACEÆ

Adoxa Moschatellina
Sambucus nigra
Viburnum opulus
,, Lantana
Lonicera Periclymenum

RUBIACEÆ

Rubia peregrina
Galium verum
,, palustre
,, uliginosum
,, saxatile
,, Mollugo
,, anglicum
,, tricorne
,, Aparine
Sherardia arvensis
Asperula odorata
,, cynanchica

VALERIANACEÆ

Valeriana dioica
,, officinalis
Valerianella olitoria

DIPSACEÆ

Dipsacus sylvestris
Scabiosa succisa
,, columbaria
,, arvensis

COMPOSITÆ.

Tragopogon minor
Helminthia echioides
Picris hieracioides
Leontodon hirtus
,, hispidus

Leontodon autumnalis
Hypochœris radicata
Lactuca virosa
,, muralis
Sonchus arvensis
,, oleraceus
Crepis virens
,, biennis
Hieracium Pilosella
,, umbellatum
,, boreale
Taraxacum officinale
Lapsana communis
Cichorium Intybus
Arctium Lappa
Carduus nutans
,, crispus
,, tenuifloris
,, lanceolatus
,, eriophorus
,, palustris
,, arvensis
,, acaulis
Onorpordum Acanthium
Carlina vulgaris
Centaurea nigra
,, Cyanus
,, Scabiosa
,, Calcitrapa
Eupatorium cannabinum
Artemisia Absinthium
,, vulgaris
Gnaphalium uliginosum
Filago germanica
Petasites vulgaris
Tussilago Farfara
Erigeron acris
Solidago Virgaurea

Senecio vulgaris
,,   viscosus
,,   erucifolius
,,   Jacobæa
,,   aquaticus
Inula Conyza
,,   dysenterica
Bellis perennis
Chrysanthemum segetum
,,   Leucanthemum
,,   Tanacetum
,,   inodorum
,,   Chamomilla
Achillea Millefolium

CAMPANULACEÆ.

Campanula rotundifolia
,,   Trachelium

JASMINACEÆ.

Fraxinus excelsior
Ligustrum vulgare

APOCYNACEÆ.

Vinca minor

GENTIANACEÆ.

Gentiana Amarella
,,   campestris
Erythræa Centaurium
,,   pulchella
Chlora perfoliata
Menyanthes trifoliata

CONVOLVULACEÆ.

Convolvulus arvensis
,,   sepium
,,   Soldanella
Cuscuta Trifolii

SOLANACEÆ.

Hyoscyamus niger

Solanum nigrum
,,   Dulcamara

SCROPHULARIACEÆ.

Verbascum Thapsus
,,   nigrum
Scrophularia nodosa
,,   aquatica
Digitalis purpurea
Linaria Cymbalaria
,,   spuria
,,   Elatine
,,   vulgaris
,,   minor
Veronica arvensis
,,   serpyllifolia
,,   scutellata
,,   Anagallis
,,   Beccabunga
,,   officinalis
,,   montana
,,   Chamædrys
,,   hederifolia
,,   agrestis
,,   Buxbaumii
Euphrasia officinalis
Bartsia Odontites
Pedicularis palustris
Rhinanthus Crista-galli
Melampyrum pratense

OROBANCHACEÆ.

Orobanche caryophyllaceæ
,,   minor
    var. amethystea
,,   Picridis
Lathræa squamaria

VERBENACEÆ.

Verbena officinalis

LABIATÆ.

Lycopus europæus
Mentha sylvestris
,, hirsuta
,, arvensis
Thymus Serpyllum
Origanum vulgare
Calamintha Acinos
,, officinalis
Nepeta Cataria
,, Glechoma
Salvia verbanaca
Prunella vulgaris
Scutellaria galericulata
,, minor
Marrubium vulgare
Ballota nigra
Stachys Betonica
,, sylvatica
,, arvensis
Galeopsis Ladanum
,, Tetrahit
Lamium Galeobdolon
,, album
,, amplexicaule
,, purpureum
,, incisum
Ajuga reptans
Teucrium Scorodonia

BORAGINACEÆ.

Echium vulgare
Lithospermum officinale
,, arvense
Myosotis palustris
,, arvensis
,, collina
,, versicolor

Anchusa sempervirens
Borago officinalis
Symphytum officinale
Cynoglossum officinale

PRIMULACEÆ.

Primula vulgaris
var. caulescens
,, veris
Lysimachia vulgaris
,, nemorum
,, nummularia
Anagallis arvensis
,, cærulea
,, tenella
Glaux maritima
Samolus Valerandi

PLUMBAGINACEÆ.

Armeria maritima
Statice binervosa

PLANTAGINACEÆ.

Plantago major
,, media
,, lanceolata
,, Coronopus

CHENOPODIACEÆ.

Suæda maritima
Beta maritima
Chenopodium olidum
,, murale
,, album
,, Bonus-Henricus
Atriplex patula

POLYGONACEÆ.

Polygonum Persicaria
,, Hydropiper
,, aviculare

Rumex Hydrolapathum
,, crispus
,, obtusifolius
,, nemorosus
,, conglomeratus
,, pulcher
,, Acetosa
,, Acetosella

ELÆAGNACEÆ.

Hippophae rhamnoides

THYMELACEÆ.

Daphne Laureola

EUPHORBIACEÆ.

Euphorbia Helioscopia
,, stricta
,, exigua
,, Peplus
,, amygdaloides
Mercurialis perennis
,, annua

URTICACEÆ.

Urtica urens
,, dioica
Parietaria officinalis
Humulus Lupulus
Ulmus montana
,, suberosa

AMENTIFERÆ.

Quercus Robur
Fagus sylvatica
Corylus Avellana
Alnus glutinosa
Populus alba
,, nigra
Salix alba

Salix caprea

CONIFERÆ.

Juniperus communis

TYPHACEÆ.

Sparganium simplex
,, ramosum
Typha latifolia
,, angustifolia

ARACEÆ.

Arum maculatum

LEMNACEÆ.

Lemna minor
,, trisulca

NAIADACEÆ.

Potamogeton densus
,, lucens
,, natans

ALISMACEÆ.

Triglochin maritimum
,, palustre
Sagittaria sagittifolia
Alisma Plantago

HYDROCHARIDACEÆ.

Hydrocharis Morsus-ranae

ORCHIDACEÆ.

Aceras anthropophora
Orchis morio
,, mascula
,, ustulata
,, pyramidalis
,, incarnata
,, maculata
Gymnadenia conopsea
Habenaria bifolia
,, chlorantha

Habenaria viridis
Ophrys apifera
,, aranifera
,, arachnites
,, muscifera
Spiranthes autumnalis
Listera ovata
Neottia Nidus-avis
Epipactis latifolia

IRIDACEÆ.

Iris fœtidissima
,, Pseudacorus

AMARYLLIDACEÆ.

Narcissus Pseudo-narcissus
Galanthus nivalis

DIOSCOREACEÆ.

Tamus communis

TRILLIACEÆ.

Paris quadrifolia

LILIACEÆ.

Polygonatum multiflorum
Convallaria majalis
Ornithogallum umbellatum
Allium vineale
         var. compactum
,, ursinum

JUNCACEÆ.

Juncus communis
,, glaucus
,, lamprocarpus
Juncus obtusiflorus
,, supinus
,, Gerardi
,, bufonius
Luzula sylvatica
,, campestris

CYPERACEÆ.

Cyperus longus
Scirpus lacustris
,, carinatus
,, setaceus
,, maritimus
,, sylvaticus
,, palustris
Eriophorum angustifolium
Carex pulicaris
,, stellulata
,, ovalis
,, remota
,, arenaria
,, divisa
,, muricata
,, divulsa
,, vulpina
,, paniculata
,, lævigata
,, sylvatica
,, pendula
,, Pseudo-cyperus
,, glauca
,, hirta
,, ampullacea
,, riparia

GRAMINA.

Anthoxanthum odoratum
Phlæum pratense
Alopecurus pratensis
,, geniculatus
,, agrestis
Agrostis canina
,, vulgaris
,, alba
Arundo Phragmites

Aria cæspitosa
,, flexuosa
,, caryophyllea
Avena pubescens
,, flavescens
Arrhenatherum avenaceum
Holcus lanatus
,, mollis
Triodia decumbens
Koeleria cristata
Melica uniflora
Molinia cærulea
Catabrosa aquatica
Glyceria fluitans
Sclerochloa rigida
Poa annua
,, pratensis
,, trivialis
Briza media
Cynosurus cristatus
Dactylis glomerata
Festuca sciuroides
,, ovina
,, elatior
var. arundinacea
Bromus giganteus
,, asper
,, sterilis
,, mollis

Brachypodium sylvaticum
,, pinnatum
Triticum caninum
,, repens
var. littorale
Lolium perenne
,, temulentum
Hordeum pratense
,, murinum
Nardus stricta
Lepturis filiformis

FILICES.

Polypodium vulgare
Polystichum aculeatum
Lastrea Filix-mas
,, spinulosa
,, dilatata
Athyrium Filix-fœmina
Asplenium Adiantum-
nigrum
,, Ruta-muraria
Scolopendrium vulgare
Blechnum boreale
Pteris aquilina
Ophioglossum vulgatum

EQUISETACEÆ.

Equisetum maximum
,, arvense
,, limosum

II.—BUTTERFLIES.

——:o:——

PAPILIONIDÆ.

Papilionidi.

Papilio Machaon

Pieridi.

Gonepteryx Rhamni
Colias Edusa
,, *var.* Helice
,, Hyale
Pieris Cratægi
,, Brassicæ
,, Rapæ
,, Napi
,, Daplidice
Anthocaris Cardamines
Leucophasia Sinapis

NYMPHALIDÆ.

Satyridi.

Arge Galathea
Satyrus Ægeria
,, Megæra
,, Semele
,, Janira
,, Tithonus
,, Hyperanthus
Chortobius Pamphilus

Vanessidi.

Vanessa Cardui
,, Atalanta
,, Io
,, Polychloros
,, Antiopa

Vanessa Urticæ
,, C-album

Argynnidi.

Argynnis Aglaia
,, Adippe
,, Lathonia
,, Euphrosyne
,, Selene
Melitæa Artemis
,, Cinxia

ERYCINIDÆ.

Nemeobius Lucina

LYCÆNIDÆ.

Thecla Rubi
,, Quercus
,, Betulæ
Polyommatus Phleas
Lycæna Ægon
,, Agestis
,, Alexis
,, Adonis
,, Corydon
,, Alsus
,, Argiolus

HESPERIDÆ.

Syrichthus Alveolus
Thanaos Tages
Hesperia Sylvanus
,, Comma
,, Linea

III.—MOTHS.

SPHINGINA.

ZYGÆNIDÆ.

Procris Statices
,, Globulariæ
,, Geryon
Zygæna Trifolii
,, Filipendulæ

SPHINGIDÆ

Smerinthus ocellatus
,, Populi
Acherontia Atropos
Sphinx Convolvuli
,, Ligustri
Deilephila Galii
,, lineata
Chærocampa Porcellus

Chærocampa Elpenor

SESIIDÆ.

Macroglossa stellatarum
,, fuciformis
,, bombyliformis
Sesia myopiformis
,, culiciformis
,, formiciformis
,, chrysidiformis
,, ichneumoniformis
,, cynipiformis
,, tipuliformis
,, andreniformis
,, bembeciformis
,, apiformis

BOMBYCINA.

HEPIALIDÆ.

Hepialus hectus
,, lupulinus
,, sylvinus
,, Velleda
,, Humuli

ZEUZERIDÆ.

Zeuzera Æsculi
Cossus ligniperda

NOTODONTIDÆ.

Dicranura furcula
,, bifida
,, vinula

Stauropus Fagi
Notodonta Camellina
,, dictæa
,, dictæoides
,, ziczac
,, chaonia
,, Dodonæa
Diloba cæruleocephala
Ptilodontis palpina
Clostera anachoreta
,, reclusa
,, curtula
Pygæra bucephala

LIPARIDÆ.

Liparis Monacha
,, dispar
,, Salicis
,, auriflua
,, chrysorrhea
Orgyia pudibunda
,, antiqua
Demas Coryli

LITHOSIDÆ.

Calligenia miniata
Lithosia mesomella
,, complanula
,, complana
,, stramineola
,, quadra
,, rubricollis
Setina irrorella
Nudaria mundana

NOLIDÆ.

Syntomis phegea
Nola cucullatella
,, cristulalis
,, albulalis
Sarrothripa revayana
Halias prasinana

CHELONIDÆ.

Callimorpha Dominula
Euthemonia russula
Chelonia caja
,, villica
,, Plantaginis
Arctia fuliginosa
,, lubricipeda
,, Menthrasti
,, Urticæ
Euchelia Jacobeæ
Deiopeia pulchella

BOMBYCIDÆ.

Bombyx neustria
,, Rubi
,, Quercus
Eriogaster lanestris
Trichiura Cratægi.
Odonestis potatoria
Lasiocampa quercifolia
Saturnia Carpini

PLATYPTERYGIDÆ.

Cilix spinula.
Platypteryx lacertula
,, falcula
,, hamula

NOCTUINA.

NOCTUO-BOMBYCIDÆ.

Thyatira derasa
,, batis
Cymatophora duplaris
,, fluctuosa
,, Or
,, flavicornis

BRYOPHILIDÆ.

Bryophila glandifera

Bryophila perla

BOMBYCOIDÆ.

Diphthera Orion
Acronycta tridens
,, Psi
,, leporina
,, Aceris
,, megacephala
,, Alni

Acronycta Ligustri
,, Rumicis
,, auricoma

LEUCANIDÆ.

Leucania conigera
,, lithargyria
,, albipuncta
,, Comma
,, straminea
,, impura
,, pallens
,, Phragmitidis
Tapinostola Bondii
Nonagria fulva
,, Hellmanni
,, geminipuncta
,, typhæ
,, lutosa

APAMIDÆ.

Gortyna flavago
Hydræcia nictitans
,, micacea
Axylia putris
Xylophasia rurea
,, lithoxylea
,, sublustris
,, polyodon
,, hepatica
,, scolopacina
Neuria Saponariæ
Heliophobus popularis
Pachetra leucophæa
Cerigo Cytherea
Luperina testacea
,, Cespitis
Mamestra abjecta
,, anceps

Mamestra furva
,, brassicæ
,, persicariæ
Apamea basilinea
,, connexa
,, gemina
,, fibrosa
,, oculea
Miana strigilis
,, fasciuncula
,, literosa
,, furuncula
,, arcuosa

CARADRINIDÆ.

Grammesia trilinea
Caradrina Morpheus
,, Alsines
,, blanda
,, cubicularis

NOCTUIDÆ.

Rusina tenebrosa
Agrotis valligera
,, puta
,, suffusa
,, saucia
,, Segetum
,, exclamationis
,, corticea
., cinerea
,, nigricans
,, Tritici
,, aquilina
,, ravida
,, lucernea
Tryphæna Ianthina
,, fimbria

Tryphæna pronuba
,,   interjecta
Noctua glareosa
,, augur
,, plecta
,, C-nigrum
,, triangulum
,, brunnea
,, festiva
,, Dahlii
,, Rubi
,, umbrosa
,, baja
,, xanthographa

ORTHOSIDÆ.

Trachea piniperda
Tæniocampa gothica
,,   leucographa
,,   rubricosa
,,   instabilis
,,   stabilis
,,   gracilis
,,   miniosa
,,   munda
,,   cruda
Orthosia suspecta
,, Ypsilon
,, lota
,, macilenta
Anchocelis rufina
,,   pistacina
,,   lunosa
,,   Litura
Cerastis Vaccinii
,, spadicea
,, erythrocephala
Scopelosoma satellitia

Oporina croceago
Xanthia cerago
,, silago
,, gilvago
,, ferruginea
Cirrhœdia xerampelina

COSMIDÆ.

Tethea retusa
Cosmia trapezina
,, affinis

HADENIDÆ.

Eremobia ochroleuca
Dianthecia carpophaga
,,   capsincola
,,   cucubali
,,   albimacula
,,   conspersa
Hecatera dysodea
,, serena
Polia flavicincta
Epunda lutulenta
,, viminalis
,, Lichenea
Miselia Oxyacanthæ
Agriopis Aprilina
Phlogophora meticulosa
Euplexia lucipara
Aplecta herbida
,, occulta
,, nebulosa
,, tincta
,, advena
Hadena Protea
,, dentina
,, Chenopodii
,, oleracea

Hadena Pisi
,, thalassina
,, contigua
,, Genistæ
Xylocampa lithorhiza
Calocampa vetusta
,, exoleta
Xylina rhizolitha
,, semibrunnea
,, petrificata
Cucullia Verbasci
,, Asteris
,, Chamomillæ
,, umbratica

HELIOTHIDÆ.

Heliothis marginata
,, peltigera
,, armigera
,, dipsacea
Heliodes Arbuti

ACONTIDÆ.

Agrophila sulphuralis
Acontia luctuosa

ERASTRIDÆ.

Erastria fuscula

PHALÆNOIDÆ.

Brephos Parthenias

PLUSIDÆ.

Habrostola Urticæ

Plusia orichalcea
,, chrysitis
,, Festucæ
,, V-aureum
,, Gamma

GONOPTERIDÆ.

Gonoptera Libatrix

AMPHIPYRIDÆ.

Amphipyra pyramidea
,, Tragopogonis
Mania typica
,, Maura

TOXOCAMPIDÆ.

Toxocampa Pastinum

STILBIDÆ.

Stilbia anomala

CATOCALIDÆ.

Catocala Fraxini
,, nupta
,, sponsa
,, promissa

OPHIUSIDÆ.

Ophiodes lunaris

EUCLIDIDÆ.

Euclidia Mi
,, glyphica

POAPHILIDÆ.

Phytometra ænea

GEOMETRINA.

URAPTERYDÆ.
Urapteryx sambucata
ENNOMIDÆ.
Epione apiciaria
Rumia cratægata

Venilia maculata
Angerona prunaria
Metrocampa margaritata
Ellopia fasciaria
Eurymene dolabraria

Pericallia syringaria
Selenia illunaria
,, lunaria
Odontoptera bidentata
Crocallis elinguaria
Ennomos tiliaria
,, fuscantaria
,, angularia
Himera pennaria

AMPHIDASIDÆ:

Phigalia pilosaria
Amphidasis prodromaria
., betularia

BOARMIDÆ.

Hemerophila abruptaria
Cleora lichenaria
Boarmia repandata
,, rhomboidaria
,, consortaria
Tephrosia consonaria
,, crepuscularia
,, biundularia
,, extersaria
,, punctulata
Gnophos obscurata

GEOMETRIDÆ.

Pseudoterpna cytisaria
Geometra papilionaria
Iodis vernaria
,, lactearia
Phorodesma bajularia
Hemithea thymiaria

EPHYRIDÆ.

Ephyra porata
,, punctaria
,, omicronaria

Ephyra pendularia

ACIDALIDÆ.

Asthena luteata
,, candidata
,, sylvata
Eupisteria heparata
Acidalia rubricata
,, scutulata
,, bisetata
,, rusticata
,, osseata
,, dilutaria
,, incanaria
,, ornata
,, promutata
,, subsericeata
,, immutata
,, remutata
,, strigilata
,, imitaria
,, aversata
,, inornata
,, emarginata
Timandra amataria

CABERIDÆ.

Cabera pusaria
,, rotundaria
,, exanthemaria
Corycia temerata
,, taminata

MACARIDÆ.

Macaria notata
,, liturata
Halia wavaria

FIDONIDÆ.

Strenia clathrata

L

Panagra petraria
Numeria pulveraria
Fidonia atomaria
,, piniaria
Minoa euphorbiata
Sterrha sacraria
Aplasta ononaria
Aspilates strigillaria
,, citraria
,, gilvaria

ZERENIDÆ.

Abraxas grossulariata
,, ulmata
Ligdia adustata
Lomaspilis marginata

HYBERNIDÆ.

Hybernia rupicapraria
,, leucophearia
,, progemmaria
,, defoliaria
Anisopteryx æscularia

LARENTIDÆ.

Cheimatobia brumata
Oporabia dilutata
Larentia didymata
,, multistrigaria
,, pectinitaria
Emmelesia affinitata
,, alchemillata
,, albulata
,, decolorata
,, unifasciata
Eupithecia venosata
,, linariata
,, centaureata
,, subfulvata
,, subumbrata

Eupithecia plumbeolata
,, isogrammata
,, satyrata
,, castigata
,, virgaureata
,, albipunctata
,, pusillata
,, pimpinellata
,, fraxinata
,, subnotata
,, vulgata
,, expallidata
,, absinthiata
,, assimilata
,, subciliata
,, abbreviata
,, exiguata
,, pumilata
,, coronata
,, rectangulata
Lobophora viretata
,, lobulata
,, polycommata
Hypsipetes impluviata
,, elutata
Melanthia ocellata
,, albicillata
,, hastata
,, procellata
,, unangulata
,, rivata
,, subtristata
,, montanata
,, galiata
,, fluctuata
Anticlea rubidata
,, badiata
,, derivata

Anticlea berberata
Coremia propugnata
,,   ferrugata
,,   unidentaria
,,   quadrifasciaria
Camptogramma bilineata
,,   fluviata
Phibalapteryx tersata
,,   lignata
,,   vitalbata
Scotosia dubitata
,,   rhamnata
,,   certata
,,   undulata
Cidaria psittacata
,,   miata
,,   picata
,,   corylata
,,   russata

Cidaria immanata
,,   suffumata
,,   silaceata
,,   prunata
,,   testata
,,   fulvata
,,   pyraliata
,,   dotata
Pelurga comitata

EUBOLIDÆ.

Eubolia cervinaria
,,   mensuraria
,,   palumbaria
,,   bipunctaria
,,   lineolata
Anaitis plagiata
Chesias spartiata
,,   obliquaria

PYRALIDINA.

DELTOIDES.

HYPENIDÆ.

Hypena proboscidalis
,,   rostralis
Hypenodes albistrigalis
,,   costæstrigalis

HERMINIDÆ.

Rivula sericealis
Herminia derivalis
,,   barbalis
,,   tarsipennalis
,,   grisealis

PYRALIDES.

ODONTIDÆ.

Odontia dentalis

PYRALIDÆ.

Pyralis fimbrialis
,,   glaucinalis
Aglossa pinguinalis

CLEDEOBIDÆ.

Cledeobia angustalis

ENNYCHIDÆ.

Pyrausta punicealis
,,   purpuralis
,,   ostrinalis
Herbula cespitalis
Ennychia cingulalis
,,   anguinalis
,,   octomaculalis

ASOPIDÆ.

Agrotera nemoralis

Endotricha flammealis

Diasemia ramburialis

Cataclysta lemnalis
Hydrocampa nymphæalis
,,   stagnalis

Botys pandalis
,,   flavalis
,,   hyalinalis
,,   verticalis
,,   lancealis
,,   fuscalis
,,   urticalis
Ebulea crocealis
,,   verbascalis
,,   sambucalis
Pionea forticalis
,,   margaritalis
,,   stramentalis
Spilodes sticticalis
,,   palealis
,,   cinctalis
Scopula olivalis
,,   prunalis
,,   ferrugalis
Lemiodes pulveralis
Mecyna polygonalis
Stenopteryx hybridalis

Scoparia ambigualis
,,   cembræ
,,   dubitalis
,,   ingratella

Scoparia lineola
,,   mercurella
,,   cratægella
,,   angustea
,,   pallida

Eromene ocellea
Crambus verellus
,,   falsellus
,,   pratellus
,,   dumetellus
,,   fascuellus
,,   pinetellus
,,   latistrius
,,   perlellus
,,   Warringtonellus
,,   tristellus
,,   inquinatellus
,,   genicunellus
,,   culmellus
,,   rorellus
,,   hortuellus
Chilo phragmitellus
Schænobius forficellus
,,   gigantellus
Ilithyia carnella
Homœosoma sinuella
,,   nimbella
,,   nebulella
,,   binævella
,,   saxicola
,,   petrella
Ephestia flutella
,,   pinguis
Cryptoblabes bistriga
Phycis betulella
,,   adornatella
,,   subornatella

Phycis ornatella
,, roborella
Rhodophæa consociella
,, advenella
,, marmorea
,, suavella
,, tumidella

Rhodophæa rubrotibiella
Oncocera ahenella
Melia sociella
,, anella
,, cephalonica
Galeria cerella
Meliphora alveariella

IV.—BIRDS.

ORDER I.—ACCIPITRES.

FAMILY—FALCONIDÆ.

Peregrine Falcon ...	Falco peregrinus
Hobby	,, subbuteo
Kestrel	,, tinnunculus
Sparrowhawk	Accipiter Nisus
Buzzard	Buteo vulgaris
Rough-legged Buzzard ...	,, lagopus
Honey Buzzard ...	Pernis apivorus

STRIGIDÆ.

Long-eared Owl ...	Otus vulgaris
Short-eared ,, ...	,, brachyotus
White ,, ...	Strix flammea
Brown ,, ...	Syrnium aluco

ORDER II.—PASSERES.

FAMILY—LANIIDÆ.

Red Backed Shrike ...	Lanius collurio

MUSCICAPIDÆ.

Spotted Flycatcher ..:	Muscicapa grisola

TURDIDÆ.

Missel Thrush	Turdus viscivorus
Fieldfare	,, pilaris
Song Thrush	,, musicus
Redwing	Turdus iliacus
Blackbird	,, merula
Ring Ouzel	,, torquatus
Golden Oriole	Oriolus galbula

SYLVIIDÆ.

Hedge Sparrow	...	Accentor modularis
Redbreast	...	Erythaca rubecula
Redstart	...	Ruticilla phænicura
Stonechat	...	Saxicola rubicola
Whinchat	...	,, rubetra
Wheater	...	,, Ænanthe
Grasshopper Warbler	...	Calamodyta locustella
Sedge ,,	...	,, phragmatis
Reed ,,	...	,, arundinacea
Nightingale	...	Philomela luscinia
Blackcap	...	Sylvia Trochilus
Garden Warbler	...	Curruca hortensis
Whitethroat	...	,, cinerea
Lesser Whitethroat	...	,, sylviella
Wood Warbler	...	,, sibilatrix
Willow ,,	...	Sylvia Trochilus
Chiffchaff	...	,, **rufa**
Gold Crest	...	Regulus cristatus
Great Tit	...	Parus major
Blue ,,	...	,, cæruleus
Cole ,,	...	,, ater
Marsh ,,	...	,, palustris
Longtailed Tit	...	,, caudatus
Pied Wagtail	...	Motacilla Yarrellii
White ,,	...	,, alba
Gray ,,	...	,, boarula
Tree Pipit	...	**Anthus** arboreus
Meadow Pipit	...	,, pratensis
Rock Pipit	...	,, petrosus

FRINGILLIDÆ.

Shore Lark	...	Alauda alpestris
Sky ,,	...	,, arvensis

Wood Lark ...	...	Alauda arborea
Snow Bunting ...	...	Plectrophanes nivalis
Common Bunting	...	Emberiza miliaris
Blackheaded Bunting	...	,, schæniclus
Yellow Bunting	...	,, citrinella
Chaffinch ...	...	Fringilla Cœlebs
Mountain Finch	...	,, montifringilla
Tree Sparrow ...	...	Passer montanus
House ,. ...	...	,, domesticus
Greenfinch ...	...	Coccothraustes chloris
Goldfinch ...	...	Carduelis elegans
Siskin ...	...	,, spinus
Linnet ..	...	Linota cannabina
Lesser Redpole	...	,, linaria
Mountain Linnet	...	,, flavirostris
Bullfinch ...	...	Pyrrhula vulgaris

STURNIDÆ.

Starling ...	...	Sturnus vulgaris

CORVIDÆ.

Raven ...	...	Corvus Corax
Carrion Crow ...	...	,, Corone
Hooded ,, ...	...	,, Cornix
Rook ...	...	,, frugilegus
Jackdaw ...	...	,, monedula
Magpie ...	...	Pica caudata
Jay ...	...	Garrulus glandarius

CERTHIADÆ.

Creeper ...	...	Certhia familiaris
Wren ...	...	Troglodytes vulgaris
Nuthatch ...	...	Sitta Europea

UPUPIDÆ.

Hoopoe ...	...	Upupa epops

ALCEDINIDÆ.

| Kingfisher | ... | ... | Alcedo ispida |

CYPSELIDÆ.

| Swift | ... | ... | Cypselus apus |

HIRUNDINIDÆ.

Chimney Swallow	...	Hirundo rustica
Martin	...	...	,, urbica
Sand Martin	...	...	,, riparia

CAPRIMULGIDÆ.

| Goatsucker | ... | ... | Caprimulgus Europeus |

ORDER III.—SCANSORES.

FAMILY—PICIDÆ.

Great Spotted Woodpecker	Picus major
Green	,,	,, viridis
Lesser	,,	,, minor
Wryneck	...	...	Yunx torquilla

CUCULIDÆ.

| Cuckoo | ... | ... | Cuculus canorus |

ORDER IV—COLUMBÆ.

FAMILY—COLUMBIDÆ.

Ring Dove	...	...	Columba palumbus
Rock ,,	...	...	,, livia
Turtle ,,	...	...	,, Turtur

ORDER V.—GALLINÆ.

FAMILY—PHASIANIDÆ.

| Pheasant | ... | ... | Phasianus colchicus |

TETRAONIDÆ.

Partridge	...	...	Perdix cinerea
Red-legged Partridge	...	,, rufa
Quail	...	...	Coturnix communis

ORDER VII.—GRALLÆ.

FAMILY—CHARADRIIDÆ.

Great Plover	Œdicnemus crepitans
Golden Plover	Charadrius pluvialis
Ringed Plover	,, hiaticula
Lapwing	Vanellus cristatus
Oystercatcher	Hæmatopus ostralegus

ARDEIDÆ.

Heron	Ardea cinerea

SCOLOPACIDÆ.

Curlew	Numenius arquata
Green Sandpiper ...	Totanus ochropus
Common ,, ...	,, hypoleucus
Woodcock	Scolopax rusticola
Common Snipe... ...	,, gallinago
Jack ,, ...	,, gallinula
Dunlin	Tringa variabilis
Purple Sandpiper ...	,, maritima
Gray Phalarope ...	Phalaropus lobatus

RALLIDÆ.

Land Rail	Crex pratensis
Baillon's Crake... ...	,, pygmæa
Water Rail	Rallus aquaticus
Moorhen	Gallinula chloropus
Coot	Fulica atra

ORDER VIII.—NATATORES.

FAMILY—ANATIDÆ

Brent Goose	Anser Brenta
Wild Swan	Cygnus ferus
Bewick's Swan ...	,, minor
Wild Duck	Anas Boschas
Teal	,, crecca

Wigeon	... ...	Anas Penelope
Velvet Duck	... ...	Oidemia fusca
Scoter	... ...	,, nigra
Goldeneye	... ...	Fuligula clangula
Redbreasted Merganser	...	Mergus serrator

COLYMBIDÆ.

Little Grebe	... ...	Podiceps minor
Blackthroated Diver	...	Colymbus arcticus
Redthroated	,, ...	,, septentrionalis

ALCIDÆ.

Guillemot	... ...	Uria troile
Little Auk	... ...	Mergulus alle
Razorbill	... ...	Alca torda

PELICANIDÆ.

Cormorant	... ...	Phalacrocorax Carbo
Gannet	... ...	Lula Bassana

LARIDÆ.

Common Tern ...	...	Sterna Hirunda
Lesser ,, ...	...	,, minuta
Black ,, ...	...	,, fissipes
Little Gull	... ...	Larus minutus
Blackheaded Gull	...	,, ridibundus
Kittiwake	... ...	,, tridactylus
Lesser Blackbacked Gull		,, fuscus
Great ,, ,,		,, marinus
Common Gull ...	...	,, canus
Herring ,, ...	...	,, argentatus
Common Skua	...	Lestris cataractes
Richardson's Skua	...	,, parasiticus

PROCELLARIDÆ.

Forktailed Petrel	...	Thalassidroma Leachii
Storm ,,	...	,, pelagica

V.—LAND AND FRESHWATER SHELLS.

AQUATIC.

CONCHIFERA.

Sphæriidæ.

Sphærium corneum
  ,,   lacustre
Pisidium amnicum
  ,,   nitidum

Unionidæ.

Unio tumidus
  ,, pictorum
Anodonta cygnea
    ,,   *var.* radiata
    ,,   ,, incrassata

NOTE.—The shells of the above Unionidæ are found in the Hythe Canal. But this is now periodically flushed with sea water, so that the molluscs have died out. I have some remarkably fine specimens of *Anodonta* (seven inches straight through), some of which have shells of the Acorn Barnacle attached. —H.U.

GASTEROPODA.

Paludinidæ.

Bythinia tentaculata
  ,,   Leachii
Hydrobia ventrosa

Limnœidæ.

Planorbis nitidus
  ,,   nautileus
  ,,   albus
  ,,   spirorbis
  ,,   vortex
  ,,   complanatus
  ,,   corneus
  ,,   contortus
Physa fontinalis
Limnæa peregra
  ,,   *var.* labiosa
  ,,   auricularia
  ,,   stagnalis
    palustris
  ,,   *var.* elongata
  ,,   truncatula
  ,,   *var.* albida
Ancylus fluviatilis

LAND AND FRESHWATER SHELLS.

TERRESTRIAL.

Helicidæ.

Succinea putris
,, elegans
Vitrina pellucida
Zonites cellarius
,, alliarius
,, nitidulus
,, radiatulus
,, nitidus
,, crystallinus
,, fulvus
Helix Aculeata
,, pomatia
,, *var.* albida
,, aspersa
,, *var.* albo-fasciata
,, ,, exalbida
,, ,, conoidea
,, nemoralis
,, *var.* hortensis
,, ,, hybrida
,, arbustorum
,, *var.* flavescens
,, ,, major
,, ,, alba
,, Cantiana
,, Cartusiana
,, *var.* rufilabris
,, rufescens
,, *var.* albida
,, ,, minor
,, concinna
,, hispida
,, virgata
,, *var.* subaperta

Helix *var.* subglobosa
,, ,, submaritima
,, ,, carinata
,, caperata
,, *var.* major
,, ,, ornata
,, ,, subscalaris
,, ,, Gigaxii
,, ericetorum
,, *var.* alba
,, ,, minor
,, rotundata
,, pygmæa
,, pullchella
,, *var.* costata
,, lapicida
Bulimus obcurus
Pupa umbilicata
,, marginata
Vertigo pygmæa
,, edentula
Balia perversa
Clausilia rugosa
,, *var.* Everetti
,, ,, tumidula
,, Rolphii
,, laminata
Cochlicopa lubrica
,, *var.* hyalina
Achatina acicula
Carychiidæ.
Carychium minimum
Cyclostomatidæ.
Cyclostoma elegans
Acme lineata

www.ingramcontent.com/pod-product-compliance
Lightning Source LLC
Chambersburg PA
CBHW021810190326
41518CB00007B/528